MATHEMATICAL HORIZONS
FOR QUANTUM PHYSICS

LECTURE NOTES SERIES
Institute for Mathematical Sciences, National University of Singapore

Series Editors: Louis H. Y. Chen and Ser Peow Tan
Institute for Mathematical Sciences
National University of Singapore

ISSN: 1793-0758

---

*For the complete list of titles in this series, please go to
http://www.worldscibooks.com/series/LNIMSNUS

Lecture Notes Series, Institute for Mathematical Sciences,
National University of Singapore

Vol.
20

MATHEMATICAL HORIZONS
FOR QUANTUM PHYSICS

Editors

Huzihiro Araki
Kyoto University, Japan

Berthold-Georg Englert
Centre for Quantum Technologies, National University of Singapore, Singapore

Leong-Chuan Kwek
Centre for Quantum Technologies, National University of Singapore, Singapore
Nanyang Technological University, Singapore

Jun Suzuki
National Institute of Informatics, Japan

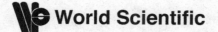 **World Scientific**

NEW JERSEY · LONDON · SINGAPORE · BEIJING · SHANGHAI · HONG KONG · TAIPEI · CHENNAI

Published by

World Scientific Publishing Co. Pte. Ltd.

5 Toh Tuck Link, Singapore 596224

USA office: 27 Warren Street, Suite 401-402, Hackensack, NJ 07601

UK office: 57 Shelton Street, Covent Garden, London WC2H 9HE

British Library Cataloguing-in-Publication Data
A catalogue record for this book is available from the British Library.

**Lecture Notes Series, Institute for Mathematical Sciences, National University of Singapore —
Vol. 20**
MATHEMATICAL HORIZONS FOR QUANTUM PHYSICS
Copyright © 2010 by World Scientific Publishing Co. Pte. Ltd.

ISBN-13 978-981-4313-31-5
ISBN-10 981-4313-31-9

Printed in Singapore.

CONTENTS

FOREWORD

The Institute for Mathematical Sciences at the National University of Singapore was established on 1 July 2000. Its mission is to foster mathematical research, both fundamental and multidisciplinary, particularly research that links mathematics to other disciplines, to nurture the growth of mathematical expertise among research scientists, to train talent for research in the mathematical sciences, and to serve as a platform for research interaction between the scientific community in Singapore and the wider international community.

The Institute organizes thematic programs which last from one month to six months. The theme or themes of a program will generally be of a multidisciplinary nature, chosen from areas at the forefront of current research in the mathematical sciences and their applications.

Generally, for each program there will be tutorial lectures followed by workshops at research level. Notes on these lectures are usually made available to the participants for their immediate benefit during the program. The main objective of the Institute's Lecture Notes Series is to bring these lectures to a wider audience. Occasionally, the Series may also include the proceedings of workshops and expository lectures organized by the Institute.

The World Scientific Publishing Company has kindly agreed to publish the Lecture Notes Series. This Volume, "Mathematical Horizons for Quantum Physics", is the twentieth of this Series. We hope that through the regular publication of these lecture notes the Institute will achieve, in part, its objective of promoting research in the mathematical sciences and their applications.

February 2010

Louis H. Y. Chen
Ser Peow Tan
Series Editors

PREFACE

Quantum theory is one of the most important intellectual developments in the early twentieth century. The mathematical theory of quantum physics emerged largely from works, pioneered by John von Neumann, on the spectral theory of linear operators on a Hilbert space in the mid 1920s, and many important developments of the last eighty years are owed to a strong interplay between theoretical physics and mathematics. Moreover, in the last twenty to thirty years, there have been phenomenal advances in which mathematicians found new tools and motivations through physical concepts and physicists exploited ideas and techniques that were originally developed for the use in pure mathematics.

It was therefore felt timely that we should raise interest in mathematical physics among researchers and students around the world, and in particular at the National University of Singapore (NUS) as well as elsewhere in Singapore, by organizing a mathematical workshop on various mathematical aspects of quantum theory. It was also felt that such a collaboration between scientists of different backgrounds, different expertise, and different scientific culture could bear fruit on the research of all participants by intellectual cross-fertilization. So, one of the major objectives of the workshop was to bring together mathematicians, whose work has a bearing on quantum physics, with researchers from mathematical and theoretical physics. This book arose from such a workshop.

The eventual theme for the workshop was "Mathematical Horizons for Quantum Physics" and the event was co-organized by the Institute for Mathematical Sciences (IMS) and the Centre for Quantum Technologies (CQT) at NUS and held on the premises of IMS over an eight-week period in 2008. In order to allow leading experts to mingle and discuss freely with young researchers and students, the workshop was organized with fewer than usual lectures and a lot of informal, interactive tutorials and discussions.

The eight-week period from July 28 to 21 September 2008 was divided into four sessions, each lasting three weeks, with an overlap of one or two weeks between successive sessions. The overlap periods gave opportunities for mutual interactions between participants from different sessions. At the end of the workshop, there were more than eighty active participants who had contributed to the discussions in one way or another.

In addition to the workshop activities, there were also two very interesting public talks: Burkhard Kümmerer spoke on "Knot or Not Knot" and Reinhard Werner delivered a talk that addressed the question "Are Quantum Computers the Next Generation of Supercomputers?" During the workshop, one of the organizers, Huzihiro Araki, also gave two presentations: the first one on the lives and careers of the Japanese Nobel Laureates Hideki Yukawa and Sin-Itiro Tomonaga at the Faculty of Science, NUS, and a second one concerning the history and mathematics of the Tomita–Takesaki theory for operator algebras at the Department of Mathematics, NUS.

During each session of the workshop, a number of pedagogical lectures, accessible to young researchers and graduate students, were provided. This book compiles the written accounts of some of these lectures. We therefore hope that this compilation will prove to be useful for graduate students and researchers who would like to start their research in an area covered in the workshop and also for researchers who require solid introductory materials and surveys of the status of the field.

The themes of the four sessions of the workshop and the respective lectures are as follows:

Session 1. **Quantum Control and Dynamics**
The central topics discussed were

- Quantum control of the alignment and orientation of polar molecules;
- Quantum chaos;
- Laser-driven models in quantum computing systems.

Arne Keller's treatment of *Control of the Molecular Alignment or Orientation by Laser Pulses* and the contribution on *Quantum Computing and Devices: A Short Introduction* by

Zhigang Zhang, Viswanath Ramakrishna, and Goong Chen as well as Hans-Rudolf Jauslin's and Dominique Sugny's account of *Dynamics of Mixed Classical-Quantum Systems, Geometric Quantization and Coherent States* are the three book chapters that originated in this session.

Session 2. **Operator Algebras in Quantum Information**
The discussions focused mostly on

- Entropy in quantum channels and the problem of additivity of quantum capacity;
- Stability of quantum algorithms in the presence of external noise;
- Entanglement of multipartite and infinite systems.

These topics are covered in another three book chapters: *Quantum Memories as Open Systems* by Robert Alicki, *Two Mathematical Problems in Quantum Information Theory* by Alexander S. Holevo, and *Dissipatively Induced Bipartitie Entanglement* by Fabio Benatti.

Session 3. **Non-Equilibrium Statistical Mechanics**
In their discussions the participants tried to answer the questions

- Is there a large deviation theory for quantum fluctuations?
- How can one construct non-equilibrium steady states?

Jan Dereziński's essay on *Scattering in Non-Relativistic Quantum Field Theory* is the book chapter for this session.

Session 4. **Strongly Interacting Many-Particle Systems**
The discussions addressed

- The theory of large atoms, molecules, and solids;
- The mathematical description of the radiation field and its interaction with matter.

Volker Bach's chapter on *Mathematical Theory of Atoms and Molecules* deals with these matters.

This volume would not have been possible without the immense efforts and contributions from the lecturers who agreed to prepare, present, and then write up their lectures at the workshop. As the organizers for the workshop, we would like to take this opportunity to thank all authors for their hard work. We are confident that this volume conveys the exciting atmosphere of all those stimulating discussions.

We would also like to thank Prof. Louis Chen, Director of IMS, A/Prof. Leung Ka Hin, the then Deputy Director of IMS, and Prof. K. K. Phua for their strong encouragement and precious advice during all stages of planning and conducting this workshop. We are grateful to the Lee Foundation and the Dean of the Faculty of Science, NUS, Prof. Andrew Wee, for their generous financial support. Without the budget contribution from the Lee Foundation, certain events — such as public talks and interaction sessions held during the workshop — would not have been possible.

We are equally thankful for the clerical support provided by the IMS secretariat, with special thanks to Agnes Wu and Claire Tan. During the workshop Stephen Auyong, IT manager at IMS, rendered invaluable support to the participants. Many others have also contributed to the success of the workshop. In particular, we would like to single out Evon Tan and Bess Fang from CQT who took splendid care of so many errands.

January 2010

Huzihiro Araki
Kyoto University, Japan

Berthold-Georg Englert
CQT and Department of Physics
National University of Singapore, Singapore

Leong-Chuan Kwek
CQT, National University of Singapore
and Nanyang Technological University, Singapore

Jun Suzuki
National Institute of Informatics, Tokyo, Japan
Editors

CONTROL OF THE MOLECULAR ALIGNMENT OR ORIENTATION BY LASER PULSES

Arne Keller

Université Paris-Sud 11
Laboratoire de Photophysique Moléculaire
Bat 210, 91405 Orsay, France
E-mail: arne.keller@u-psud.fr

We review the motivations and recent theoretical and experimental progress in laser alignment or orientation of molecules in space. We present a method to produce field free aligned or oriented linear molecules using a succession of short laser pulses. The problem of orienting molecules is an example of a unitary control problem in a Hilbert space with infinite dimension. The originality of our method relies on a precise construction of oriented quantum target states which maximize the orientation in finite dimensional subspaces. Orientation at zero temperature (pure state control) and of molecular thermal ensemble (mixed state control) are investigated.

1. Introduction — Motivations

The problem of controlling the alignment or orientation of molecules in space has attracted a fair amount of attention in the field of molecular physics and chemical physics. In this introduction, we present the motivations for such studies and, and give some account of the progress.

First of all, we must define what we call molecular alignment and orientation. In classical terms, aligning a molecular axis in space means to bring an axis fixed in the molecular frame to a given fixed direction in the laboratory frame. Orienting the molecule means that in addition to aligning, we require that the orientation of the axis be fixed in the laboratory frame. This last definition makes sense only if the molecule is not symmetric with respect to a reflection on a plane orthogonal to the molecular axis we like to orient. For instance, aligning a $A - B$ linear molecule to a vertical axis (in the laboratory frame) means to bring the axis joining the two atoms A

1

and B to a vertical position. Orienting the molecule means that in addition we require for instance that A is up and B down. In the quantum world we are only allowed to speak in the sense of probability. We can define an angular probability distribution $P(\theta, \phi)$ where (θ, ϕ) are the spherical angles (co-latitude and azimuthal angle, respectively) locating the molecular axis with respect to the fixed laboratory frame axis. $P(\theta, \phi)d\Omega$ gives the probability to find the molecular axis in the $d\Omega = \sin\theta d\theta d\phi$ solid angle centered on the (θ, ϕ) position. We can consider a molecule as aligned if its angular probability distribution is mainly supported near $\theta = 0$ and/or $\theta = \pi$. If in addition $P(\theta, \phi)$ is mainly supported in the upper space only ($\theta \in [0, \frac{\pi}{2}]$ and $\phi \in [0, 2\pi]$ for instance), we consider the molecule to be oriented.

1.1. *Motivations*

Why molecular alignment and orientation are important? In the following, three examples in different areas of investigation are presented where the control of the molecular alignment and/or orientations is a crucial point.

(1) **Stereochemistry.** Stereochemistry is the study of the influence of the relative orientation of atoms and molecules on the outcomes of a chemical reactions. To our knowledge, this is the first motivation to produce aligned or oriented molecule. Indeed, this idea goes back to the 1960s, when the first molecular beam machines were built, to produces inelastic or reactive collisions. The development of the field of molecular beam chemistry and reaction dynamics in the 1960s [1] invited direct experimental study of the steric effect. The objective of those experiments is to control the relative orientation of the reagent and to measure its influence on the reaction probability. Several strategies have been applied to produces oriented or aligned beams of molecules whose presentation is deferred to the next section.

(2) **Femtosecond optics.** Nowadays the production of laser pulse with a duration of about 10-100 femtosecond (1 fs $= 10^{-15}$) can be achieved almost routinely in laboratory. Researchers would like to control precisely the amplitude and phase of such pulses for practical applications and also to compress them to sub-femtosecond durations. One line of investigation is based on the propagation of the laser pulse in a medium of oriented molecules. The main idea is that carefully controlled molecular rotational dynamics induces periodic change in the refractive index of the propagation media, then inducing controlled phase and amplitude variation of the laser pulse [2, 3].

(3) **High harmonic generation and molecular ionization.** When an atom or a molecule is irradiated with an intense (intensity about $10^{14} - 10^{15}$ W/cm^2) infrared (wavelength $\simeq 800$ nm) laser pulse, the molecule can be ionized and can also emit coherent X-UV attosecond light pulses. The accurate understanding of these processes requires the alignment or orientation of the irradiated molecule. Furthermore, recent experiments are exploring the possibility of reconstructing the initial quantum electronic molecular state (molecular state tomography) by observing the emitted light or the angular distribution of emitted electrons [4–8]. Those experiments need carefully aligned or oriented molecules.

1.2. *Experimental methods*

In this section, we give a brief survey of the methods that have been used in laboratory to align or orient molecules. The first experiment with oriented molecules was realized in 1965 by Kramer and Bernstein [1]. They used an axially symmetric hexapole electrostatic field to focus molecular beams. Symmetric top-like molecules traveling through the inhomogeneous electric field of the hexapole follow sinusoidal trajectories and focus at a certain point, depending on their rotational quantum state and the voltage on the hexapole rods (see [9] for a detailed description of the technique). This technique allows to obtain molecular beams in a given rotational state which can be oriented. In fact, in these types of experiments, it is the molecule angular momentum which is directly oriented and not the molecular axis.

Next, in the 1990s a "brute force orientation technique" was introduced by Loesh and Remscheid [10] and independently by Friedrich and Herschbach [11]. This technique made it possible to carry out experiments on oriented molecules not being symmetric tops as was necessary for applying hexapole state selection method. This technique relies on the application of an intense electrostatic field on rotationally very cold polar molecules obtained in supersonic beams. The anisotropy of the Stark effect allows molecules in the lowest few rotational states to be trapped in "pendular states" and thereby confined to librate (oscillate about the field axis) over a limited angular range.

In 1995, Friedrich and Hershbach [12] showed that an intense and non resonant laser field is able to align a molecule, with respect to the polarization of the laser electric vector, through the interaction between the oscillating laser field and the induced molecular dipole moment. The induced dipole moment results from the interaction of the laser field through the molecular anisotropic polarizability. The induced dipole moment interaction can be described by a potential energy term $V(\theta) \propto \cos^2 \theta$ where θ is the angle between the molecular axis and the laser electric field vector. This potential presents a symmetric double-well with minima at $\theta = 0$ and $\theta = \pi$. Because of the fast oscillations of the electric field this technique cannot orient the molecule; it only produces alignment. The lower energy levels correspond to the so called pendular aligned states. Later in 1999, Friedrich and Hershbach proposed [13] to combine an electrostatic field with the laser field to achieve molecular orientation in addition to alignment. The pendular level induced by the laser field comes in nearly degenerate tunneling doublet states of opposite parity. If the molecule is polar, the introduction of a static electric field couples the two components of a doublet, inducing tunneling and thus orientation. This idea has been implemented by Sakai *et al.* recently [14]. At the same time another scheme was proposed by Dion *et al.* [15] and refined by Guerin *et al.* [16]. It consists in using two lasers with different frequencies ω and 2ω and different phases. An experimental confirmation was performed by Ohmura *et al.* in 2004 [17].

All those schemes rely on a slow switching of the laser, inducing an adiabatic passage from a pure rotational state of the free molecule to an aligned or oriented pendular state of the molecule interacting with the field. The alignment or orientation can then be maintained while the molecule is interacting with the laser field. But the alignment and orientation is completely destroyed when the laser field is switched off adiabatically. This has the drawback that the intense laser field can modify the physics and/or the chemistry and can constitutes an obstacle in some applications which require free oriented molecules.

Hence, researchers have begun to focus on pulsed schemes which is the opposite of the adiabatic case, the objective being to produce free (non interacting) aligned or oriented molecules during sufficiently long time. The first experimental demonstration of post pulse alignment was achieved by Rosca-Pruna and Vrakking [18], and was followed by several other groups (for a review see [19]). In these experiments a short laser pulse interacts through the molecular polarizability, and creates a coherent superposition of rotational eigenstates. Constructive interference of the component in the wave packet, and thus alignment, occur shortly after the pulse and repeat periodically in time. The molecular alignment can be detected by a sudden dissociation of the molecule, in that case the angular distribution of the fragments reveals the alignment of the parent molecule. The molecular alignment can also be detected by non destructive spectroscopic methods [20, 21]. These non-adiabatic schemes are interesting for applications of aligned molecules because it enables alignment under field free conditions. For alignment based on a single short laser pulse, stronger alignment of a molecular sample is obtained by increasing the pulse intensity. This approach is limited by saturation of the alignment itself, or by the requirement that alignment pulse intensity must be kept below a threshold corresponding to unwanted processes such as dissociation or ionization of the molecule. The idea to use a train of laser pulses to control the molecular alignment was first proposed by Averbukh and Arvieu [22, 23]. The goal is to obatain better alignment or orientation efficiency with a lower laser intensity in each pulse.

In this article we will attempt to summarize recent results concerning the control of orientation or alignment of linear molecule by a train of laser pulses. These results have been obtained in the collaboration between the author and O. Atabek, D. Daems, C. M. Dion, S. Guerin, H. Jauslin and D. Sugny [24–27]. The chapter is organized as follows: In section 2, we present the model used to describe the molecule and its interaction with the laser pulses. Section 3 is devoted to the control problem. We first try to formalize the controls objectives and show how to build target quantum states that corresponds to an aligned or oriented molecule. Then, we show strategies to reach these target states with a train of laser pulses. Two cases are envisaged: the control of a pure state, where the molecule can be considered initially in the ground states (zero temperature). And the control of a mixed state, where the molecule is initially in a mixed state describing a thermal ensemble. Finally, we end with a conclusion and some prospects.

2. The Model

2.1. *The free molecule*

We consider a linear molecule as a rigid rotor. This is of course an approximation, because a molecule posses many other degree of freedom than the rotational one: electronic motion and vibrational motion which correspond in general to higher frequency modes. These other degrees of freedom can be safely ignored if we can consider that the molecule remains in its ground electronic an vibrational quantum state. In our case, this implies that the laser frequency used for the control is not equal to a transition energy between theses modes.

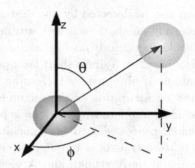

Fig. 1.   Spherical angles defining the rotor position in space.

The position of the linear molecule in space is given by the spherical angles (θ,ϕ) where θ is the co-latitude and ϕ the azimuthal angle (see Fig. 1). The Hamiltonian of the free molecule corresponds to the rotational kinetic energy:

$$H_0 = BJ^2 \tag{2.1}$$

where B is the rotational constant which is related to the inertia moment I of the molecule by $B = \frac{1}{2I}$. $J^2 = J_x^2 + J_y^2 + J_z^2$, where $J_i (i = x, y, z)$ are the components in the laboratory frame of the angular momentum operator. The eigenvectors $|j, m\rangle$ of $H = BJ^2$ can be chosen as eigenstates of J_z with eigenvalues m (H_0 does not depend on ϕ). We have :

$$J^2|j, m\rangle = j(j + 1)|j, m\rangle,$$
$$J_z|j, m\rangle = m|j, m\rangle, \tag{2.2}$$

with $j \in \mathbb{N}$ and $-j \le m \le +j$. To a rotor energy $E_j = Bj(j + 1)$ corresponds $2j + 1$ degenerated quantum states. In the (θ,ϕ) representation the eigenvectors are represented by spherical harmonics: $\langle\theta, \phi|j, m\rangle = Y_{j,m}(\theta, \phi)$ and the Hamiltonian is the spherical Laplacian:

$$H_0 = B\left[\frac{1}{\sin\theta}\frac{\partial}{\partial\theta}\sin\theta\frac{\partial}{\partial\theta} + \frac{1}{\sin^2\theta}\frac{\partial^2}{\partial\phi^2}\right], \tag{2.3}$$

defined on the Hilbert space $\mathcal{H} = L^2(\mathbb{S}^2)$.

The evolution operator $U_0(t, t_0)$, solution of the Schrödinger equation:

$$i\frac{\partial}{\partial t}U_0(t, t_0) = H_0U_0(t, t_0); \quad U(t_0, t_0) = \mathbb{1}, \tag{2.4}$$

which is periodic in t with period $T_{\text{rot}} = \frac{\pi}{B}$. T_{rot} is called the rotational period.

2.2. *Molecule-laser interaction*

2.2.1. *Dipole moment interaction*

If the molecule is not symmetric, it can interact with light through its permanent dipole moment $\vec{\mu}$. In that case the interaction Hamiltonian is given by:

$$V_\mu = -\vec{\mu} \cdot \vec{E}(t) = \mu E(t) \cos\theta, \qquad (2.5)$$

where $\vec{E}(t) = E(t)\vec{e}_z$ is the laser electric field which we consider as linearly polarized in the z direction. The laser electric field is written as

$$E(t) = \mathcal{E}(t) \cos\omega t, \qquad (2.6)$$

where ω is the laser angular frequency and $\mathcal{E}(t)$ is the pulse envelope with a typical duration τ. The propagation of a light pulse in vacuum requires that

$$\int_{-\infty}^{\infty} E(t)dt = 0. \qquad (2.7)$$

In a typical laser pulse, the duration τ is much longer than the carrier period $T_L = \frac{2\pi}{\omega}$, as a consequence the electric field is almost symmetric in space. It is clear that such a laser pulse cannot orient a molecule. It has been shown [28, 29] that it is possible to generate very special pulses, called half-cycle pulses, with a duration of the order of picosecond (10^{-15} s), with a very highly non symmetric shape. This pulse shape can be divided into two parts: The first part of the pulse is short and the amplitude of the electric field can reach a large value. The time integral of the electric field corresponding to this first part is not zero giving us the possibility to orient the molecule. On the contrary, the second part of the pulse takes a longer time but with a very low electric field, such that Eq. (2.7) is fulfilled. The effect of this second part of the pulse on the molecule can be safely neglected [30].

2.2.2. *Polarizability interaction*

If the molecule is symmetric (a homonuclear diatomic molecule for instance) it cannot posses a permanent dipole moment. But the interaction with the laser field can polarize the molecular electronic cloud giving rise to an induced dipole moment: $\vec{\mu}(t) = \overleftrightarrow{\alpha}\vec{E}(t)$ where $\overleftrightarrow{\alpha}$ is the molecular polarizability second rank tensor. For a linear molecule the induced $\vec{\mu}$ can be decomposed into the sum of a component parallel to the molecular axe:

$\mu_\parallel = \alpha_\parallel E_\parallel$ and an orthogonal component: $\mu_\perp = \alpha_\perp E_\perp$. The interaction of the laser field with this induced dipole moment gives the interaction potential: $V(t) = -\vec{\mu}(t) \cdot \vec{E}(t) = -(\overleftrightarrow{\alpha} \vec{E}(t)) \cdot \vec{E}(t)$. For a linear molecule, it becomes: $V(t) = -\alpha E^2(t) \cos^2 \theta - \alpha_\perp E^2(t)$, where $\alpha = \alpha_\parallel - \alpha_\perp$. In the case where the laser pulse duration is long compared to the carrier period T_L [31, 32] and to the rotational period $T_{\rm rot}$, an effective interaction operator can be obtained by averaging $V(t)$ over the period T_L. In this way we obtain the interaction Hamiltonian:

$$V_\alpha(t) = -\alpha \mathcal{E}^2(t) \cos^2 \theta - \alpha_\perp \mathcal{E}^2(t), \qquad (2.8)$$

where only the laser pulse envelope plays a role. For a symmetric molecule, only alignment makes sense, it is clear that such interaction in principle can also align the molecule.

The time-dependent Hamiltonian describing the molecule interacting with the laser field can be written as:

$$H(t) = H_0 + V_{\mu(\alpha)}(t). \qquad (2.9)$$

2.3. *Molecular states time dependent evolution*

The evolution operator for the interacting molecule is the solution of the Schrödinger equation:

$$\frac{\partial}{\partial t} U(t, t_0) = \left[H_0 + V_{\mu(\alpha)}(t) \right] U(t, t_0); \quad U(t_0, t_0) = \mathbb{1}.$$

We note that the Hamiltonian (and the evolution operator) does not depend on ϕ thus the quantum number m is conserved during the evolution.

We now consider two different cases depending on the initial molecular state.

2.3.1. *Pure state*

We first consider the case where the molecular initial state is a pure state, for instance if the temperature is low enough to ensure that only the ground rotational state of the molecule is populated. In this case the initial state is an element $|\psi(t_0)\rangle \in \mathcal{H}$ and its evolution from time t_0 to t is given by

$$|\psi(t)\rangle = U(t, t_0)|\psi(t_0)\rangle. \qquad (2.10)$$

If O is an observable, *i.e* a self adjoint operator on $\mathcal{H}$, then we define the expectation $\langle O \rangle(t)$ of O in the state $|\psi(t)\rangle$ at time t as

$$\langle O \rangle(t) = \langle \psi(t)|O\psi(t)\rangle. \qquad (2.11)$$

2.3.2. *Mixed state*

For most molecules it requires very low temperature (less than 1 K) to ensure that only the ground rotational state is populated. We thus consider also the case where the initial state of the molecule is in a statistical superposition of rotational states. In that case, the initial state at time t_0 can be described by the density operator $\rho(t_0)$ which is a self adjoint positive operator on $\mathcal{H}$, with $\text{tr}[\rho(t_0)] = 1$. For an initial thermal ensemble at temperature T the density operator is given by

$$\rho(t_0) = \left(\text{tr}\left[e^{-\beta H_0}\right]\right)^{-1} e^{-\beta H_0}, \tag{2.12}$$

where $\beta = \frac{1}{kT}$ and k is the Boltzmann constant.

The time evolution of ρ from time t_0 to time t is given by

$$\rho(t) = U(t,t_0)\rho(t_0)U^{-1}(t,t_0), \tag{2.13}$$

we see that the evolution is unitary. We note that the spectrum of $\rho(t)$ is conserved during such an evolution, and that an initial mixed state remains mixed during the evolution; it cannot be converted to a pure state.

The expectation $\langle O \rangle(t)$ of an observable O in state $\rho(t)$ at time t is given by

$$\langle O \rangle(t) = \text{tr}[\rho(t)O]. \tag{2.14}$$

2.3.3. *Sudden approximation*

We consider short laser pulses, such that the pulse duration τ is very small compared to the rotational period T_{rot}. This condition is routinely achieved in laboratory. Indeed typical molecular rotational periods are of the order of several picoseconds and laser durations less than 100 femtoseconds are now easily achievable. In that case, it can be shown [33, 34] that the evolution operator $U(t,t_0)$, solution of Eq. (2.3), can be factored as

$$U(t > \tau, 0) \simeq e^{-iH_0 t} e^{i\mathcal{V}_{\mu(\alpha)}}$$
$$= U_0(t,0)U_V,$$

where we have defined

$$\mathcal{V}_{\mu(\alpha)} = \int_{-\infty}^{+\infty} V_{\mu(\alpha)}(t')dt', \tag{2.15}$$

more precisely,

$$\mathcal{V}_\alpha = \left[\alpha \int_{-\infty}^{+\infty} \mathcal{E}^2(t')dt'\right] \cos^2\theta \equiv \mathcal{A}_\alpha \cos^2\theta,$$

$$\mathcal{V}_\mu = \left[\mu \int_{-\infty}^{+\infty} \mathcal{E}(t')\cos\omega t' dt'\right] \cos\theta \equiv \mathcal{A}_\mu \cos\theta. \tag{2.16}$$

2.4. *Alignment or orientation characterization*

To quantify the degree of molecular orientation or alignment two measures are used.

2.4.1. *Orientation*

The molecular orientation will be characterized by the expectation $\langle\cos\theta\rangle$. A perfect orientation is achieved if $|\langle\cos\theta\rangle| = 1$. Here $\cos\theta$ is considered as an operator on $\mathcal{H}$, and its matrix representation in the basis of free molecular eigenstates $|j, m\rangle$ is tridiagonal. It can be written as

$$\cos\theta = \sum_{j,|m|\leq j} C_{j,m}\left[|j+1,m\rangle\langle j,m| + |j,m\rangle\langle j+1,m|\right], \tag{2.17}$$

with $C_{j,m} = \left[\frac{(j+1)^2 - m^2}{(2j+1)(2j+3)}\right]^{\frac{1}{2}}$. We note that free molecular eigenstates $|j, m\rangle$ are not oriented: $\langle j, m| \cos\theta|j, m\rangle = 0$.

2.4.2. *Alignment*

To characterize the degree of molecular alignment, we use the expectation $\langle\cos^2\theta\rangle$. For free molecular rotational states $\langle j, m| \cos^2\theta|j, m\rangle \neq 0$ in particular $\langle j = 0, m = 0| \cos^2\theta|j = 0, m = 0\rangle = \frac{1}{3}$.

3. Control

3.1. *Control objectives*

Our goal is to maximize the orientation or the alignment of the molecule, more precisely we want to find a laser field such that the expectation $\langle\cos\theta\rangle(t_f)$ or $\langle\cos^2\theta\rangle(t_f)$ reaches its maximum value for some finite time t_f. In addition, we impose that at time t_f when the orientation (or alignment) reaches its maximum value the molecule has to be free, that is $E(t_f) = 0$. This condition ensures that experiments using such oriented molecules will not be perturbed by the orienting laser field. Furthermore, we would like to obtain an oriented (or aligned) molecule during long enough time. The maximum achieved value for the orientation (or alignment) will be called *efficiency* of the control. The time during which the orientation (alignment) takes high enough value will be called *persistence*.

To show that indeed a control is necessary, we first compute the evolution of the orientation $\langle \cos\theta \rangle(t)$ as a function of time t, after the interaction with a single short pulse. The results of such a computation are presented in

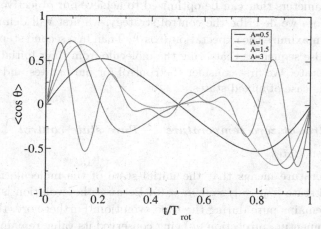

Fig. 2. $\langle \cos\theta \rangle$ as a function of time in units of the rotational T_{rot} period, for 4 values of the amplitudes of the radiative coupling $\mathcal{A}_\mu = 0.5, 1, 1.5, 3$. The initial state is the ground state $|j = 0, m = 0\rangle$ and a single pulse interacts with the molecule at time $t = 0$.

Fig. 2, for four values of the radiative coupling $\mathcal{A}_\mu$ (see Eq. (2.16)). We note that increasing $\mathcal{A}_\mu$ first increases the maximum value reached by $\langle \cos\theta \rangle$, but value of $\mathcal{A}_\mu$ higher than 1.5 does not increase any longer the value of $\langle \cos\theta \rangle$. Orientation higher than $\langle \cos\theta \rangle = 0.7$ cannot be reached with such an intense single pulse. Furthermore, the increase of the radiative coupling causes a decrease of the persistence. Indeed, if we take as a measure of the persistence the time during which the value of $\langle \cos\theta \rangle$ is higher than 0.5, we see that increasing $\mathcal{A}_\mu$ from 1 to 3 decreases the persistence by a factor of 3. This can be explained, by the fact that increasing the radiative coupling increases the kinetic energy transfered to the molecule, inducing fast rotation of the molecule.

In conclusion, to obtain good orientation during long enough time, a single intense pulse is not sufficient. An alternative idea is to use a sequence of short pulses and to optimize the delay and intensities of such pulses.

In the sudden approximation the evolution operator corresponding to a sequence of N short laser pulses at time $t_n (n = 0, 1, 2, \ldots, N-1)$ can be written as

$$U(t > t_{N-1}, 0) = e^{-iH_0(t-t_N)} \prod_{n=0}^{N-1} e^{iH_0(t_{n+1}-t_n)} e^{i\mathcal{A}_n \cos\theta}. \qquad (3.1)$$

We have at hand, $N-1$ time delays $(t_{n+1} - t_n)$ and N amplitudes $\mathcal{A}_n$, as control parameters that can be optimized to achieve our objective.

But before we describe the control strategy, we first will define target states that maximize the expectation $\langle \cos\theta \rangle$. Then in a second step we look for laser pulse sequences that bring the molecule from their initial state to the target state. We first consider the control for pure states and then we envisage the case of mixed states.

3.2. *Control at zero temperature — Pure state control*

3.2.1. *Target states*

Zero temperature means that the initial state of the molecule is simply the ground state $|\psi_0\rangle = |j = 0, m = 0\rangle$. Because the evolution is unitary, the state remains pure during the time evolution. Furthermore, the z-axis angular momentum projection m being conserved its value remains $m = 0$. Thus, we look for a state $|\chi\rangle \in \mathcal{H}$ which maximizes $\langle \chi| \cos\theta\chi \rangle$. But the maximum value of $\langle \cos\theta \rangle$ is 1, and the state that realizes this bound is not physically acceptable. It is not a smooth function of θ, or in other words it corresponds to a molecule with an infinite kinetic rotational energy.

To define physically acceptable target states we proceed as follows. We define the finite dimensional Hilbert $\mathcal{H}^{(j_{\max})}$ as the complex linear span of $\{|j, m = 0\rangle; j = 0, 1, 2, \ldots, j_{\max}\}$. We look for the state $|\chi^{(j_{\max})}\rangle \in \mathcal{H}^{(j_{\max})}$ which maximizes $\langle \chi^{(j_{\max})}| \cos\theta\chi^{(j_{\max})} \rangle$. This state, which we call the *target state* is simply the eigenvector of $\cos\theta^{(j_{\max})}$ with the highest eigenvalue $\lambda^{(j_{\max})}$, where $\cos\theta^{(j_{\max})}$ is the restriction of the operator $\cos\theta$ in the $\mathcal{H}^{(j_{\max})}$ finite Hilbert space. Of course this target state depends on the chosen dimension $j_{\max}$ for the $\mathcal{H}^{(j_{\max})}$ Hilbert space. We can choose the value of $j_{\max}$ so that the target state corresponds to a suitable oriented molecule. In Fig. 3, we present the angular probability distribution $P(\theta, \phi)$ defined as

$$P(\theta, \phi) = \left| \langle \theta, \phi | \chi^{(j_{\max})} \rangle \right|^2 = \left| \sum_{j=0}^{j_{\max}} \langle j, m = 0 | \chi^{(j_{\max})} \rangle Y_{j,0}(\theta, \phi) \right|^2, \qquad (3.2)$$

as a function of θ for three values of $j_{\max}$, $j_{\max} = 1, 3$ and 5. A good surprise is that a very well oriented state is obtained with low value of $j_{\max}$. Indeed, the orientation expectation values $\langle \cos^{(j_{\max})} \theta \rangle$ obtained in each case are: $\langle \cos^{(j_{\max})} \theta \rangle = 0.57, 0.86$ and 0.96, respectively.

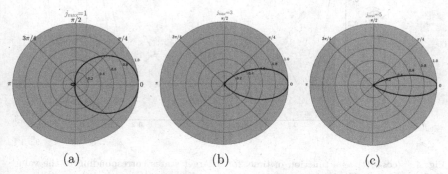

Fig. 3. Angular probability density $P(\theta, \phi = 0)/P(0)$ in polar coordinate for the target state (see Eq (3.2)) as a function of θ, for three value of $j_{\max}$. (a) $j_{\max} = 1$, (b) $j_{\max} = 3$ and (c) $j_{\max} = 5$.

To evaluate the orientation persistence of these target states we have systematically calculated the free evolution of these target state $|\chi^{(j_{\max})}(t)\rangle = U_0(t, 0)|\chi^{(j_{\max})}\rangle$ and computed their orientation expectation $\langle \cos \theta \rangle(t) = \langle \chi^{(j_{\max})}(t)| \cos \theta \chi^{(j_{\max})}(t) \rangle$ as a function of time. In Fig. 4 the time evolution of $\langle \cos \theta \rangle(t)$ is presented for $j_{\max} = 1, 3, 5$. As we have noted previously, we see that as $j_{\max}$ increases the persistence decreases. But for $j_{\max} = 5$ the orientation remains greater than $\langle \cos \theta \rangle = 0.5$ during more than than 10% of the rotational period, this can correspond to several picoseconds for a usual molecule. In Fig. 5, we present the target states efficiency and persistence for values of $j_{\max}$ ranging from $j_{\max} = 1$ to $j_{\max} = 12$. The persistence is defined as the time during which the orientation remains greater than $\langle \cos \theta \rangle = 0.5$. We see clearly in this figure that while $j_{\max}$ increases, the orientation increases but the persistence decreases. To choose the value of $j_{\max}$, a compromise must be made. We choose $j_{\max} = 5$ as a good compromise. The same method for building target states for the alignment control purpose can be applied, by replacing $\cos \theta$ by $\cos^2 \theta$. We are confident that the same approach can also be applied to other control problems in an infinite dimensional Hilbert space, where the goal is to maximize the expectation of an observable. We emphasize that this way of determining the target state does not depend on the characteristics of the molecule.

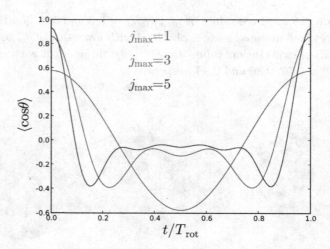

Fig. 4. $\langle \cos \theta \rangle(t)$ as a function of time, for 3 target states corresponding to the value of $j_{\max} = 1, 2, 3$. The initial state is the corresponding target state and the evolution is the free evolution computed with $U_0(t, 0)$.

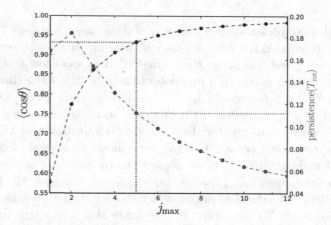

Fig. 5. Target states orientation $\langle \cos \theta \rangle$ (in blue and left vertical axis) and persistence (in red and right vertical axis) as a function of $j_{\max}$. The persistence is defined as the time during which the orientation remains greater than $\langle \cos \theta \rangle = 0.5$, and is given in unit of the rotational period T_{rot}.

Now that we have precisely determined the target state, we want to find a control strategy to reach this target state form the molecular ground state.

3.2.2. *Open questions*

Before we discuss any details of the strategies used to control the orientation, we present some open questions that are both of mathematical and physical interest. At first sight, in our context, the control objective can be formulated as:

For a given target $|\chi^{(j_{\max})}\rangle$, is it possible to find two finite sequences, τ_n and $\mathcal{A}_n$, such that

$$\left| \left\langle \chi^{(j_{\max})} \middle| \prod_{n=0}^{N-1} e^{-iH_0\tau_n} e^{i\mathcal{A}_n \cos\theta} \middle| j=0 \right\rangle \right| = 1.$$

This might be too demanding. Indeed, the infinite dimensionality of the Hilbert space $\mathcal{H}$ makes it very difficult to achieve such a task. We can reformulate the objective in the following less demanding way:

For a given target state $|\chi^{(j_{\max})}\rangle$ and for a given ϵ, are there two finite sequences, τ_n and $\mathcal{A}_n$, such that

$$1 - \left| \left\langle \chi^{(j_{\max})} \middle| \prod_{n=0}^{N-1} e^{-iH_0\tau_n} e^{i\mathcal{A}_n \cos\theta} \middle| j=0, m=0 \right\rangle \right| < \epsilon.$$

If the answer to this question is positive, then we can further ask: how many kicks are needed to reach the target state within ϵ.

Because, we do not have answer to theses questions we must proceed by doing numerical experiments.

3.2.3. *Strategy of maxima*

This strategy consists in applying a laser pulse each time $\left| \langle \chi^{(j_{\max}=5)} | \psi(t) \rangle \right|$ reaches a global maximum over a rotational period. In Fig. 6, we present the evolution of $\left| \langle \chi^{(j_{\max}=5)} | \psi(t) \rangle \right|^2$ as a function of time t, under such strategy. We see that after ten laser pulses the target state is almost reached. The target state is not exactly reached because of the infinite dimensional nature of the Hilbert space. To evaluate the robustness of this strategy, we have made calculations with different laser amplitudes $\mathcal{A}_o \pm 20\%$ but with the same time delays. The numerical results are presented in Fig. 7, showing that this strategy is indeed robust.

3.2.4. *Optimization with an evolutionary algorithm*

For experimental applications, the control of more than 2 or 3 pulses becomes a difficult task. Hence, we look for a solution with only two laser

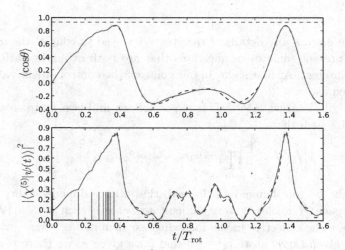

Fig. 6.   Strategy of maxima. Bottom panel: $\left|\langle\chi^{(j\max=5)}|\psi(t)\rangle\right|^2$ as a function of time t in unit of rotational period T_{rot}. Top panel: $\langle\cos\theta\rangle(t)$; the horizontal dashed line represent the value of $\langle\chi^{(j\max=5)}|\cos\theta\chi^{(j\max=5)}\rangle$. The dashed line corresponds to numerical calculation in the 6 dimensional Hilbert space $\mathcal{H}^{(j\max=5)}$. The solid line corresponds to calculation in a higher dimensional space. The initial state is $|j=0, m=0\rangle$. A pulse is applied each time $\left|\langle\chi^{(j\max=5)}|\psi(t)\rangle\right|$ reaches a global maximum over a rotational period. The laser amplitude $\mathcal{A}_o = 1$ for all pulses and ten pulses are used.

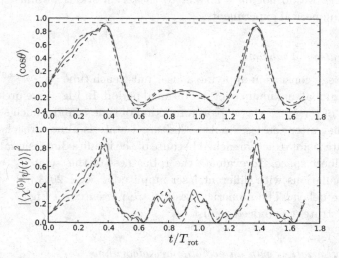

Fig. 7.   Robustness. Same calculation as Fig. 6 but with different laser amplitudes $\mathcal{A}_o$. Solid green line same as Fig. 6, $\mathcal{A}_o = 1$. Dashed red line $\mathcal{A}_o = 1.2$. Dashed blue line $\mathcal{A}_o = 0.8$.

pulses, and optimize the delay and laser amplitudes in order to maximize $\sup_{t\in[0,T_{\mathrm{rot}}]}\left|\langle\chi^{(j_{\max}=5)}|\psi(t)\rangle\right|^2$. The optimization was done with an evolutionary algorithm. The result is presented in Fig. 8, which shows that a very satisfactory solution is obtained. Indeed, the target state is almost reached, the maximum value being $\left|\langle\chi^{(j_{\max}=5)}|\psi(t)\rangle\right|^2 = 0.9887$ with pulses amplitudes given by $\mathcal{A}_o = 0.9741, 3.2930$ and a time delay between the two pulses equal to $0.2419 \times T_{\mathrm{rot}}$. The orientation reaches almost its maximal value in $\mathcal{H}^{(j_{\max}=5)}$ Hilbert space. In conclusion, we have successfully found

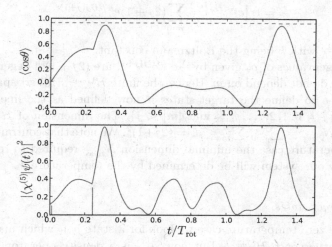

Fig. 8. Time evolution of the orientation (top) and $\left|\langle\chi^{(j_{\max}=5)}|\psi(t)\rangle\right|^2$ (bottom) with two laser pulses, where time delay and amplitudes has been optimized. The horizontal line represent the highest orientation value reachable in the $\mathcal{H}^{(j_{\max}=5)}$ Hilbert space.

a very simple solution to the problem of orienting a linear molecule at zero temperature, with only two laser pulses. The steps to find this solution was:

- Define a finite dimensional Hilbert space where the dynamic is approximated.
- Define a target state space which maximize our objective in this finite dimensional Hilbert space.
- Find a strategy to reach the target state from the initial state.

The problem is that in current laboratory experiments, the temperature is higher than the rotational energy, hence molecules cannot be considered as being initially in their ground state. In the next section we will see how to adapt the present scheme for a non zero temperature.

3.3. *Control of a thermal ensemble — Mixed state control*

We now consider a thermal ensemble initially described by the following density operator ρ_0:

$$\rho_0 = \frac{1}{Z} e^{-\beta H_0} = \frac{1}{Z} \sum_{j=0}^{\infty} \sum_{m=-j}^{+j} e^{-\beta B j(j+1)} |j, m\rangle \langle j, m|, \qquad (3.3)$$

where

$$Z = \text{tr}\left[e^{-\beta H_0}\right] = \sum_{j=0}^{\infty} (2j+1) e^{-\beta B j(j+1)} \qquad (3.4)$$

and $\beta = \frac{1}{kT}$ with k being the Boltzmann constant.

The eigenvalues of ρ_0 given by $\frac{1}{Z} e^{-\beta B j(j+1)}$ are $(2j+1)^{\text{th}}$ degenerated, since they do not depend on m. Hence, the finite $\mathcal{H}^{(j_{\max})}$ Hilbert space that we will use to define the target states is now defined as the linear span of $\{|j, m\rangle; j = 0, 1, 2, \ldots, j_{\max}$ and $|m| < j\}$. The dimension of $\mathcal{H}^{(j_{\max})}$ is given by: $N_{j_{\max}} = \sum_{j=0}^{j_{\max}} \sum_{m=-j}^{j} 1 = (2j+1)^2$. We note that contrary to the zero temperature case the minimal dimension $N_{j_{\max}}$ required to represent accurately our system will be determined by the temperature T.

3.3.1. *Target states*

As for the zero temperature case we look for a state ρ_{opt} which maximizes $\langle \cos \theta \rangle$ in the space $\mathcal{H}^{(j_{\max})}$. But now ρ_{opt} is a density operator and the expectation $\langle \cos \theta \rangle$ is given by $\langle \cos \theta \rangle = \text{tr}[\rho_{\text{opt}} \cos \theta]$. We also require that ρ_{opt} must be reachable from the initial state ρ_0. This implies that ρ_{opt} must be unitarily equivalent to ρ_0; *i.e* there exists a unitary operator U in $\mathcal{H}^{(j_{\max})}$ such that $\rho_{\text{opt}} = U^{-1} \rho_0 U$. Hence, ρ_{opt} has the same spectrum as ρ_0. The density operator ρ_{opt} fulfilling these properties is represented by a diagonal matrix in the eigenbasis of the $\cos \theta^{(j_{\max})}$ operator, and can be written as [35]

$$\rho_{\text{opt}} = \sum_{n=1}^{N_{j_{\max}}} |\chi_n\rangle \omega_n \langle \chi_n|, \qquad (3.5)$$

with $\omega_1 \leq \omega_2 \leq \cdots \leq \omega_N$ being the eigenvalues of ρ_0 ($\omega_n = \frac{1}{Z} e^{-\beta B j_n (j_n+1)}$) ordered in the same way as the eigenvalues $\chi_1 \leq \chi_2 \leq \cdots \leq \chi_N$ corresponding to eigenvectors $|\chi_n\rangle$ of $\cos \theta^{(j_{\max})}$.

One problem remains, i.e., ρ_{opt} is in fact not reachable from ρ_0 with the dynamics induced by the $U(t, t_0)$ evolution operator, indeed $U(t, t_0)$ cannot change the quantum number m. We have taken care about the kinematical constraint $\rho_{\text{opt}} = U^{-1}\rho_0 U$, but we have not taken into account the dynamical constraint coming from the specific chosen dynamics, which involves linear polarized laser pulses. To take into account this constraint, we split the $\mathcal{H}^{(j_{\max})}$ Hilbert space as $\mathcal{H}^{(j_{\max})} \doteq \oplus_{m=-j_{\max}}^{m=j_{\max}} \mathcal{H}_m^{(j_{\max})}$ where each $\mathcal{H}_m^{(j_{\max})}$ is the linear span of $\{|j, m\rangle ; j = |m|, |m|+1, |m|+2, \ldots, j_{\max}\}$. Correspondingly, the target state can be written as $\rho_t^{(j_{\max})} = \oplus_{m=-j_{\max}}^{m=j_{\max}} \rho_m^{(j_{\max})}$ where each $\rho_m^{(j_{\max})}$ is a density operator defined in the same way as ρ_{opt} by Eq. (3.5) but in $\mathcal{H}_m^{(j_{\max})}$. Physically, we may say that ρ_{opt} corresponds to an optimal state where the laser polarization has not been specified and ρ_t corresponds to the target state that can be reached with a linear polarized laser.

As for the zero temperature case we can calculate the efficiency and persistence of theses target states as a function of $j_{\max}$ (see Fig. 5). The results for a temperature $T = 5$ K and a rotational constant B such that $\beta B = 5$ which corresponds to the LiCl molecule, are presented in Fig. 9. In this figure, we see that contrary to the zero temperature case, higher $j_{\max}$'s are needed to achieve a good orientation, and consequently with a lower persistence. Nevertheless, with $j_{\max} = 6$ we obtain $\langle \cos \theta \rangle = 0.7$ and a

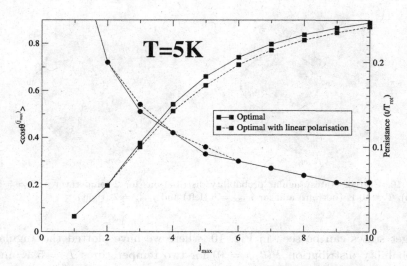

Fig. 9. Orientation (squares and left vertical axis) and persistence (circles and right vertical axis) in unit of rotational period as a function of $j_{\max}$ for the optimal target states ρ_{opt} (solid line) and the target states ρ_t (dashed line) corresponding to a linear polarized laser. The temperature is $T = 5$ K and $\frac{B}{kT} \simeq 0.2$.

persistence of $0.08T_{\text{rot}}$ which is satisfactory. Another surprising result is that the efficiency and persistence corresponding to ρ_{opt} and ρ_t are almost the same. This implies that considering linear polarized laser pulses is enough for our purpose; we cannot achieve better orientation with anymore complex polarization scheme. We emphasize that this result has been obtained without any dynamical simulation, but just by a detailed characterization of the target states. The effect of the temperature on the orientation of the

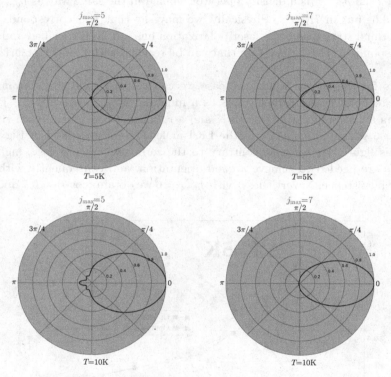

Fig. 10. Target states angular probability distributions for 2 temperatures $T = 5$ K (top), $T = 10$ K (bottom) and for $j_{\text{max}} = 5$ (left) and $j_{\text{max}} = 7$ (right).

target states can be seen in Fig. 10, where we have plotted the angular probability distribution $P(\theta, \phi = 0)$ for two temperatures $T = 5$ K and $T = 10$ K, $P(\theta, \phi)$ being defined as:

$$P(\theta, \phi) = \sum_{jm, j'm'} Y_{jm}^*(\theta, \phi) Y_{j'm'}(\theta, \phi) \langle j, m | \rho_t | j', m' \rangle . \qquad (3.6)$$

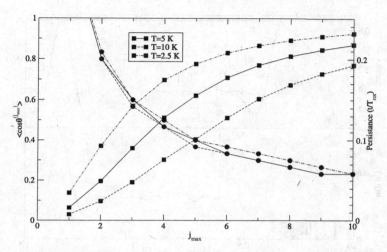

Fig. 11. Efficiency (square and left axis) and persistence (circle and right axis) in unit of $T_{\rm rot}$ for the target states ρ_t as a function of $j_{\max}$ for three temperatures: $T = 2.5$ K (dashed line), $T = 5$ K (solid line) and $T = 10$ K (dashed-solid line). We see here that as in the case of zero temperature, the efficiency increases and the persistence decreases when $j_{\max}$ increases. Furthermore, we note that the efficiency decreases with the temperature but that the persistence, if defined as the time during which the orientation remains higher than half its maximal value, does not depends on the temperature.

We have studied numerically the efficiency and persistence of target states as a function of $j_{\max}$ and temperature, for a molecule like LiCl where $B \simeq 1$ K. The results are presented in Fig. 11. They show that, as in the case of zero temperature, the orientation increases and the persistence decreases when $j_{\max}$ increase. In addition, we see that the orientation decreases when the temperature increases but the persistence almost does not depend on the temperature. Here the persistence has been defined as the time during which the orientation remains at half of its maximum value.

We have to concede that at high temperature it is nearly impossible to orient a thermal ensemble of linear molecules under unitary evolution; low temperature is required. The possibility to implement non unitary evolution (dissipative evolution) will be discussed in the conclusion section. Nevertheless, for a molecule such as LiCl at temperature $T = 5$ K ($\beta B = 0.2$), which is feasible in laboratory using a molecular supersonic beam, we can choose as a good compromise between orientation and persistence the target state corresponding to $j_{\max} = 6$. Indeed, this gives us an orientation $\langle \cos\theta \rangle \simeq 0.7$ and a persistence of $0.08 \times T_{\rm rot}$.

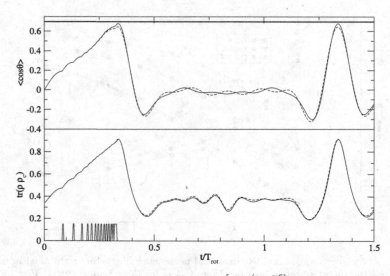

Fig. 12. Strategy of maxima. $\langle \cos \theta \rangle$ (top) and $\frac{\mathrm{tr}[\rho(t)\rho_c^{j_{max}=6}]}{\|\rho_c)\|^2}$ (bottom) as a function of time in units of the rotational period T_{rot}. The thick horizontal line indicate the optimal value $\langle \cos \theta \rangle$ in $\mathcal{H}^{j_{max}=6}$. Instants at which short laser pulses occur are indicated in the bottom panel. The temperature is $T = 5$ K. The molecule is LiCl with a rotational constant such that $\beta B \simeq 0.2$. The dashed line corresponds to an evolution calculated in the restricted Hilbert space $\mathcal{H}^{(j_{max}=6)}$ and the solid line is for an evolution in a higher dimensional Hilbert space. We have taken care of comparing the time propagation in the restricted Hilbert space $\mathcal{H}^{(j_{max})}$ and in a higher dimensional Hilbert space. The two evolutions are very similar thus showing that states outside $\mathcal{H}^{(j_{max})}$ plays only a minor role.

3.3.2. *Strategy of maxima*

We apply the same strategy of maxima that we have used in the zero temperature case. The distance between the system actual state $\rho(t)$ and the target state ρ_c is evaluated using the Frobenius norm $\|A\| = \sqrt{\mathrm{tr}[A^\dagger A]}$, where $A^\dagger$ is the adjoint of A. $\|\rho\|^2$ is also called purity of the system when ρ is a density matrix. Because under unitary evolution the purity is conserved, it is equivalent to consider the "overlap" $S(t) = \frac{\mathrm{tr}[\rho(t)\rho_c]}{\|\rho_c)\|^2}$ as a measure of distance between $\rho(t)$ and ρ_c. Hence, we apply the strategy which consists in applying a laser pulse each time the free evolution of $S(t)$ reaches a global maximum over a rotational period. The results of such strategy are shown in Fig. 12, for a temperature of $T = 5$ K for the LiCl molecule and with the target state corresponding to $j_{max} = 6$. We see that the target state is almost reached with 15 pulses with equal amplitudes $\mathcal{A}_o = 1$.

4. Conclusion

We have presented a general method to control the alignment or orientation of a molecule by successive laser pulses. This control problem corresponds to a unitary control of pure or mixed states in an infinite dimensional Hilbert space. The originality of the method relies on a precise definition of target states which represent the objectives of the control. These target states are defined in a finite dimensional Hilbert space. As a consequence, the controlled dynamics that bring the initial state to one target state must restrict approximatively the system evolution in a finite dimensional space, with good accuracy. We think that this works with molecular rotational states, mainly because the energy spacing increases linearly when the state energy increases. It would be interesting to investigate if the method presented here can be applied to other systems or other degrees of freedom, which do not posses such a property.

The method presented here to orient molecule relies on the use of terahertz coherent pulses, the so called half cycle pulses [28, 29]. Until now, no experimental implementation of this scheme has been realized, which might be due to the difficulty to produce and manipulate such pulses. Recently, Sugarawa *et al.* [36] have proposed a method to orient molecule which combines the adiabatic and the pulsed method. The oriented state is reached by switching on adiabatically the laser field in the presence of a static field, but the laser is then switched off suddenly (in less than 200 fs). This method allows to obtain field-free oriented molecules. This theoretical proposal have been demonstrated experimentally by Goban *et al.* [37].

We have presented only the unitary control problem. We have shown that even with complex polarization scheme, very cold molecular ensembles are needed to obtains well oriented molecules. This is due to the fact that thermal mixed oriented target states are less oriented than pure target states, and that the unitary evolution cannot transform a pure state to a mixed state. It is natural then to investigate the non-unitary control problem. Physically, we can think of a molecule interacting with a gas buffer or in a liquid. The first attempt to address the molecular alignment or orientation dissipative control is due to Ramakrishna and Seideman [38–40]. The generalization of the method presented here to build target states in the case of dissipative control has been done by D. Sugny *et al.* [41]. The conclusion is that the presence of the dissipative medium is an obstacle to the orientation. It is not possible to purify the molecular state if the interaction with the dissipative medium is not controlled.

References

1. K. H. Kramer and R. B. Bernstein. Focusing and orientation of symmetric-top molecules with the electric six-pole field. *J. Chem. Phys.*, 15:767, 1965.
2. R. A. Bartels, T. C. Weinacht, N. Wagner, M. Baertschy, Chris H. Greene, M. M. Murnane, and H. C. Kapteyn. Phase modulation of ultrashort light pulses using molecular rotational wave packets. *Phys. Rev. Lett.*, 88(1):013903, Dec 2001.
3. V. P. Kalosha and J. Herrmann. Phase relations, quasicontinuous spectra and subfemtosecond pulses in high-order stimulated raman scattering with short-pulse excitation. *Phys. Rev. Lett.*, 85(6):1226–1229, Aug 2000.
4. R. Velotta, N. Hay, M. B. Mason, M. Castillejo, and J. P. Marangos. High-order harmonic generation in aligned molecules. *Phys. Rev. Lett.*, 87(18):183901, Oct 2001.
5. R. de Nalda, E. Heesel, M. Lein, N. Hay, R. Velotta, E. Springate, M. Castillejo, and J. P. Marangos. Role of orbital symmetry in high-order harmonic generation from aligned molecules. *Phys. Rev. A*, 69(3):031804, Mar 2004.
6. D. Zeidler, H. Niikura, H. Pépin, J. C. Kieffer, P. B. Corkum, D. M. Villeneuve, J. Itatani, and J. Levesque. Tomographic imaging of molecular orbitals. *Nature*, 432:867, 2004.
7. D. PaviCic, K. F. Lee, D. M. Rayner, P. B. Corkum, and D. M. Villeneuve. Direct measurement of the angular dependence of ionization for N_2, O_2, and CO_2 in intense laser fields. *Phys. Rev. Lett.*, 98(24):243001, 2007.
8. H. Merdji, P. Breger, G. Waters, M. Stankiewicz, L. J. Frasinski, R. Taieb, J. Caillat, A. Maquet, P. Monchicourt, B. Carre, P. Salieres, W. Boutu, and S. Haessler. Coherent control of attosecond emission from aligned molecules. *Nature*, 4:545, 2008.
9. R. W Anderson. Tracks of symmetric top molecules in hexapole electric fields. *J. Phys. Chem.*, 101:7664, 1997.
10. A. Remscheid and H. J. Loesch. Brute force in molecular reaction dynamics: A novel technique for measu ring steric effects. *J. Chem. Phys.*, 93(7):4779–4790, 1990.
11. H. J. Loesh, J. Bulthuis, and J. Moller. Spatial orientation of molecules in strong electric fields and evidence for pendular states. *Nature*, 353:412, 1991.
12. B. Friedrich and D. Herschbach. Alignment and trapping of molecules in intense laser fields. *Phys. Rev. Lett.*, 74(23):4623–4626, Jun 1995.
13. B. Friedrich and D. Herschbach. Enhanced orientation of polar molecules by combined electrostatic and nonresonant induced dipole forces. *J. Chem. Phys.*, 111(14):6157–6160, 1999.
14. H. Sakai, S. Minemoto, H. Nanjo, H. Tanji, and T. Suzuki. Controlling the orientation of polar molecules with combined electrostatic and pulsed, nonresonant laser fields. *Phys. Rev. Lett.*, 90(8):083001, Feb 2003.
15. O. Atabek, A. Keller, H. Umeda, C. M. Dion, A. D. Bandrauka, and Y. Fujimura. Two-frequency ir laser orientation of polar molecules. numerical simulations for hcn. *Chem. Phys Lett.*, 302:215, 1999.

16. S. Guérin, L. P. Yatsenko, H. R. Jauslin, O. Faucher, and B. Lavorel. Orientation of polar molecules by laser induced adiabatic passage. *Phys. Rev. Lett.*, 88(23):233601, May 2002.

17. H. Ohmura and T. Nakanaga. Quantum control of molecular orientation by two-color laser fields. *J. Chem. Phys.*, 120(11):5176–5180, 2004.

18. F. Rosca-Pruna and M. J. J. Vrakking. Experimental observation of revival structures in picosecond laser-induced alignment of $i2$. *Phys. Rev. Lett.*, 87(15):153902, Sep 2001.

19. H. Stapelfeldt and T. Seideman. Colloquium: Aligning molecules with strong laser pulses. *Rev. Mod. Phys.*, 75(2):543–557, Apr 2003.

20. V. Renard, M. Renard, S. Guérin, Y. T. Pashayan, B. Lavorel, O. Faucher, and H. R. Jauslin. Postpulse molecular alignment measured by a weak field polarization technique. *Phys. Rev. Lett.*, 90(15):153601, Apr 2003.

21. V. Renard, M. Renard, A. Rouzée, S. Guérin, H. R. Jauslin, B. Lavorel, and O. Faucher. Nonintrusive monitoring and quantitative analysis of strong laser-field-induced impulsive alignment. *Phys. Rev. A*, 70(3):033420, Sep 2004.

22. I. Sh. Averbukh and R. Arvieu. Angular focusing, squeezing, and rainbow formation in a strongly driven quantum rotor. *Phys. Rev. Lett.*, 87(16):163601, Sep 2001.

23. M. Leibscher, I. Sh. Averbukh, and H. Rabitz. Molecular alignment by trains of short laser pulses. *Phys. Rev. Lett.*, 90(21):213001, May 2003.

24. D. Sugny, A. Keller, O. Atabek, D. Daems, C. M. Dion, S. Guérin, and H. R. Jauslin. Reaching optimally oriented molecular states by laser kicks. *Phys. Rev. A*, 69(3):033402, Mar 2004.

25. D. Sugny, A. Keller, O. Atabek, D. Daems, C. M. Dion, S. Guérin, and H. R. Jauslin. Laser control for the optimal evolution of pure quantum states. *Phys. Rev. A*, 71(6):063402, 2005.

26. D. Sugny, A. Keller, O. Atabek, D. Daems, C. M. Dion, S. Guérin, and H. R. Jauslin. Control of mixed-state quantum systems by a train of short pulses. *Phys. Rev. A*, 72(3):032704, 2005.

27. D. Sugny, A. Keller, O. Atabek, D. Daems, C. M. Dion, S. Guérin, and H. R. Jauslin. Control of mixed-state quantum systems by a train of short pulses. *Phys. Rev. A*, 72(3):032704, 2005.

28. B. I. Greene, J. F. Federici, D. R. Dykaar, R. R. Jones, and P. H. Bucksbaum. Interferometric characterization of 160 fs far-infrared light pulses. *Appl. Phys. Lett.*, 59(8):893–895, 1991.

29. R. R. Jones, D. You, and P. H. Bucksbaum. Ionization of Rydberg atoms by subpicosecond half-cycle electromagnetic pulses. *Phys. Rev. Lett.*, 70(9):1236–1239, Mar 1993.

30. O. Atabek C. M. Dion, and A. Keller. Orienting molecules using half-cycle pulses. *Eur. J. Phys. D*, 14:249, 2001.

31. A. Keller. Alignment of linear molecules with an intense ir laser: a rigid rotor high frequecy floquet treatment. *Journal of Molecular Structure (Theochem)*, 493:103, 1999.

32. A. Keller, C. M. Dion, and O. Atabek. Laser-induced molecular rotational dynamics: A high-frequency floquet approach. *Phys. Rev. A*, 61(2):023409, Jan 2000.

33. M. Machholm and N. E. Henriksen. Field-free orientation of molecules. *Phys. Rev. Lett.*, 87(19):193001, Oct 2001.

34. D. Sugny, A. Keller, O. Atabek, D. Daems, S. Guérin, and H. R. Jauslin. Time-dependent unitary perturbation theory for intense laser-driven molecular orientation. *Phys. Rev. A*, 69(4):043407, Apr 2004.

35. M. D. Girardeau, S. G. Schirmer, J. V. Leahy, and R. M. Koch. Kinematical bounds on optimization of observables for quantum systems. *Phys. Rev. A*, 58(4):2684–2689, Oct 1998.

36. Y. Sugawara, A. Goban, S. Minemoto, and H. Sakai. Laser-field-free molecular orientation with combined electrostatic and rapidly-turned-off laser fields. *Phys. Rev. A*, 77(3):031403, 2008.

37. A. Goban, S. Minemoto, and H. Sakai. Laser-field-free molecular orientation. *Phys. Rev. Lett.*, 101(1):013001, 2008.

38. S. Ramakrishna and T. Seideman. Intense laser alignment in dissipative media as a route to solvent dynamics. *Phys. Rev. Lett.*, 95(11):113001, 2005.

39. S. Ramakrishna and T. Seideman. Dissipative dynamics of laser induced nonadiabatic molecular alignment. *J. Chem. Phys.*, 124(3):034101, 2006.

40. A. Pelzer, S. Ramakrishna, and T. Seideman. Optimal control of molecular alignment in dissipative media. *J. Chem. Phys.*, 126(3):034503, 2007.

41. D. Sugny, C. Kontz, and H. R. Jauslin. Target states and control of molecular alignment in a dissipative medium. *Phys. Rev. A*, 74(5):053411, 2006.

QUANTUM COMPUTING AND DEVICES:
A SHORT INTRODUCTION

Zhigang Zhang

Department of Mathematics
University of Houston
Houston, TX 77004, USA
E-mail: zgzhang@math.uh.edu

Viswanath Ramakrishna

Department of Mathematical Sciences
University of Texas at Dallas
Richardson, TX 75080, USA
E-mail: vish@utdallas.edu

Goong Chen

Department of Mathematics and Institute for Quantum Studies
Texas A&M University
College Station, TX 77843, USA
and
Taida Institute for Mathematical Sciences
National Taiwan University
Taipei, Taiwan, ROC
E-mail: gchen@math.tamu.edu

Quantum computation is a new interdisciplinary research enterprise involving physicists, mathematicians, computer scientists and engineers. In this chapter, we give a quick introduction of the subject. Three fundamental principles of quantum informatics: superposition, entanglement and reversibility, are described first. The quantum devices of cavity-QED and optical lattices, as candidates for quantum computing gates are introduced next. Finally, we offer an algebraic treatment by conjugacy to obtain some results on universal quantum gates by explicit conjugation.

1. Introduction

The design and construction of the future quantum computer is an exciting development of the twenty-first century. This is a new interdisciplinary field, involving physicists, mathematicians, computer scientists and engineers. Here, we will attempt to give a short course of the basics of quantum computing, with somewhat more emphasis on the mathematical aspects. Much of the material in this chapter is scattered but available in the literature.

Two book sources on quantum computing are:

(i) A general introduction to quantum computation and quantum information by Nielsen and Chuang [28], and
(ii) A book with emphasis on quantum computing devices, by Chen, Church, Englert, Henkel, Rohwedder, Scully and Zubairy [10].

In recent years, more books on quantum computing have been published, at a rapid rate and too many for us to mention here. Also, the number of research articles appear at an ever accelerating speed. No matter how well intentioned and planned we (or any other authors) wish to present an up-to-date introductory or survey paper on the topic of quantum computing, we must profess that our knowledge is actually quite limited. Nevertheless, we do hope that this article serves some useful purpose for the interested readership. The organization of the chapter is made as follows:

(1) In Section 2, we explain three of the fundamental principles of quantum information science: *superposition*, *entanglement* and *reversibility*, following essentially [10] and [11].
(2) In Section 3, we introduce the mathematical modeling of quantum computing devices. Some recent development has been included.
(3) Section 4 contains original material. Partial success of a unifying treatment for universality of quantum gates has been achieved.

2. A Brief Historical Account, Motivation, and Three Fundamental Principles of Quantum Computing

Quantum information science has a relatively short history. The celebrated physicist R. P. Feynman talked about simulating physics with computers [18] in 1982. Actually two years earlier, in 1980, P. Benioff, another physicist at the Argonne National Lab near Chicago, already published an article [2] introducing a quantum Turing machine model. Further, in the mid and late 1980s, D. Deutsch published his papers [14, 15] formulating concrete approaches and introducing a quantum circuit model.

This topic of quantum computing has begun to grow and attract researchers' attention. But a major breakthrough, in the form of a "killer app," due to the mathematician P. Shor, then at the Bell Labs, published his quantum factoring algorithm during the mid-1990s [36]. It has really generated tremendous enthusiasm in the scientific community for quantum computing and make the field blossom. Shor's algorithm utilizes quantum computers' peculiar properties of superposition and entanglement in quantum mechanics to achieve massive parallelism and unprecedented speedup. Such properties do not have any classical analogues and, thus, quantum computers have the potential to execute certain special tasks "exponentially faster" than classical computers.

On the hardware front, the computer industry will soon be facing one of its greatest challenges — the end of *Moore's Law*. G. E. Moore, a co-founder of Intel, observed the empirical law that the number of transistors that can be built into an integrated circuit doubled every 18 months and at half the cost. (Actually, he said 24 months instead of 18 months.) Such efforts to miniaturize microchips in electronics, if continue to be successful, will allow nine or ten more doublings and hit a brick wall in 20 to 25 years. The transistors built into the chip will be of the size of several atoms, and the current semi-conductor technology will not permit the size to be reduced further. Microelectronics is inevitably moving into nanoelectronics. Therefore, the development of quantum information science and quantum computing technology most surely constitutes the future of information science and technology. It has also become the major impetus to the development of new general quantum technologies, which, since the invention of the laser during the 1950s, have changed nearly every aspect of life for the modern human being.

From the historical perspective, the infrastructure and foundation for quantum computing were built more than half a century earlier. The following three profound events actually constitute the most important preludes that paved the way for the development of a modern quantum computer (QC):

(1) The Stern-Gerlach experiment (1920s);
(2) The Einstein-Podolsky-Rosen (EPR) paper [17] (1935) and its reply from Schrödinger [31];
(3) The Landauer principle on information erasure [25] (1961).

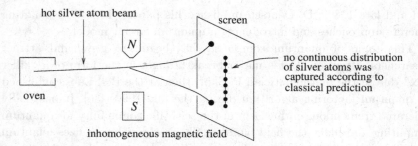

Fig. 1.   A beam of spin-1/2 particles is sent through an inhomogeneous magnetic field. The magnetic field is strongly increasing in one direction; let's call this privileged "z" direction. Each such particle has a spin magnetic moment. The inhomogeneous magnetic field causes the particles to be deflected, 1/2 of them up and 1/2 of them down, totally separated rather than in a continuous distribution, and thus measures the magnetic moment of the particle and shows that the spin is discrete and (spatially) quantized.

2.1. *Spins, the Stern-Gerlach experiment and superposition*

Spin refers to intrinsic angular momentum possessed by particles such as individual atoms, protons, or electrons. It is particularly important in quantum mechanics for systems at atomic length scales or smaller.

The fact that electrons have spins was first discovered in the celebrated Stern-Gerlach experiment, named after the German physicists Otto Stern and Walther Gerlach, in 1922 on deflection of silver atoms. The story can be read in an interesting recent account by Friedrich and Herschbach [19], where it was pointed out that if it were not due to the cheap cigars that the two poor burgeoning physicists Stern and Gerlach were smoking, the split silver atom beams hitting the screen might never have been recorded.

The general concept of elementary particle spin was first proposed in 1925 by R. Kronig, G. Uhlenbeck, and S. Goudsmit [20, 21, 38]. Electrons are "spin-1/2" particles because their intrinsic spin angular momentum has $s = 1/2$. Other elementary spin-1/2 particles include neutrinos and quarks. On the other hand, photons are spin-1 particles, i.e., $s = 1$.

The physical setup of the Stern-Gerlach experiment as shown in Figure 1 will be called a Stern-Gerlach apparatus (SGA) from now on. The particles whose trajectories are in the upper and lower split beams are said to have spin-up and spin-down, denoted by $|\uparrow\rangle$ and $|\downarrow\rangle$, respectively.

Thus, the SGA demonstrates a basic, intrinsically quantum property of spins of elementary particles. (Nevertheless, we need to say that there is no Stern-Gerlach experiment for electrons or protons because they carry electric charges and the Lorentz force will deflect them differently.)

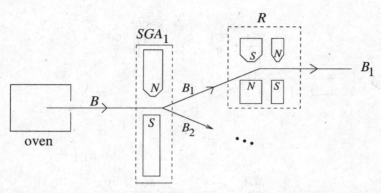

Fig. 2. A beam B of atoms is split by an SGA_1, resulting in two split beams B_1 and B_2. Now let the upper beam B_1 pass through a "direction rectifier" R. The function of R is to straighten the direction of B_1 to that of the original B.

Now, let us modify the SGA slightly, so that a split beam can propagate along the same direction (parallel to that) of the original beam; see Figure 2.

From Figure 2, we see that we have an exit beam B_1 propagating along a direction parallel to that of the original beam B. The intensity of B_1 is $1/2$ of the original B. Now, consider three possible actions on the beam B_1:

(i) Apply to B_1 an SGA'', aligned in exactly the same direction as SGA_1; see Figure 3(a).

(ii) Apply to B_1 an $SGA^\perp$ aligned in the direction perpendicular to SGA_1; see Figure 3(b).

(iii) Apply to B_1 an SGA^θ aligned in a direction forming an angle θ with respect to the direction of alignment of SGA_1; see Figure 3(c).

If we write the spin-up $|\uparrow\rangle$ and spin-down $|\downarrow\rangle$ states, respectively, as *quantum bit* (*qubit*) states $|0\rangle$ and $|1\rangle$. Then the experiment in Figure 1 indicates that atoms in the beam are representable in the form of a linear combination, i.e., *superposition*

$$\frac{1}{\sqrt{2}}(|0\rangle + |1\rangle),$$

where the factor $1/\sqrt{2}$ is a normalization factor in order to achieve a unit L^2-norm (as $|0\rangle$ and $|1\rangle$ are two orthogonal states).

Indeed, the experiments in Figure 3(c) indicate that a more general superposition can be of the form

$$\frac{1}{\sqrt{2}}\left(\cos\frac{\theta}{2}|0\rangle + e^{i\phi}\sin\frac{\theta}{2}|1\rangle\right), \tag{2.1}$$

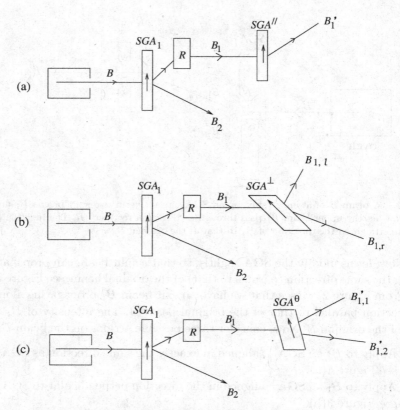

Fig. 3.   Consider the exit beam B_1 from the configuration as in Figure 2. We now let B_1 pass another SGA as set up in the upper right of (a), (b), (c) above:

(a) For SGA$''$ whose alignment is identical to SGA$_1$, the beam B_1 is deflected upward to B_1', and the beam does not split.

(b) For SGA$^\perp$, the beam B_1 splits into a left beam $B_{1,\ell}$ and a right beam $B_{1,r}$, with the same intensity.

(c) For SGA$^\theta$, the beam splits into two beams $B_{1,1}'$ and $B_{1,2}'$; one of them has intensity $\cos^2 \frac{\theta}{2}$, the other has $\sin^2 \frac{\theta}{2}$, where θ is the angle of the alignment of SGA$^\theta$ relative to that of SGA$_1$.

i.e., spin states $|0\rangle$ and $|1\rangle$ can have different weights $\cos^2 \frac{\theta}{2}$ and $\sin^2 \frac{\theta}{2}$ as well as a phase factor $e^{i\phi}$ where the phase factor ϕ makes the superposition *complex*. The appearance of ϕ is not obvious from the experiments in Figure 3. However, we can point out that in many natural quantum system operations or evolutions, a so-called *Rabi oscillation* naturally occurs, which will make a quantum state take a complex form such as in Equation (2.1).

The above discussions applying SGAs to a particle beam can be reduced to the case where only a *single* particle (e.g., a silver atom) is emitted from the particle gun. Then what we observe is that the single particle has spin-up with probability 1/2, and spin-down with probability 1/2. This can be done experimentally by using detectors catching and then counting the particles in the directions of $B'_{1,1}$ and $B'_{1,2}$. As a matter of fact, in general depending on how this particle was "prepared," Equation (2.1) shows that the probabilities of spin-up can range from 0% to 100%.

2.2. *Entanglement and the Einstein-Podolsky-Rosen (EPR) Problem*

A much more complicated situation arises when we deal with two particle spins or other types of arrangements when they are *correlated*. This is the important entanglement case, a scenario arising, e.g., in an atomic decay problem where two correlated particles are emitted almost simultaneously.

In 1935, Einstein, Podolsky and Rosen (EPR) [17] drew on this special quantum entanglement phenomenon to show that measurements performed on spatially separated parts of a quantum system can have an instantaneous effect on each other, constituting a paradoxical so-called "nonlocal behavior". Indeed, EPR's elegant presentation of the entanglement (Schrödinger's word) is the basis for the Bell inequality, as well as much of the stuff of modern quantum mechanics such as *quantum teleportation* and *quantum dense coding*, etc.

We set the stage as follows. Consider a 2-atom molecule (called a *dimer*) such as H_2, i.e., the hydrogen molecule consisting of two hydrogen atoms, with a combined zero angular momentum of spin. We know that the hydrogen atom has a single electron (with a nucleus consisting of a single proton). In the dimer H_2, because of the *Pauli exclusion principle*, the combined spins of the two electrons in H_2 (in its ground state) must be *antisymmetric*, in a so-called *spin-singlet* configuration:

$$\frac{1}{\sqrt{2}}(|\uparrow_1\downarrow_2\rangle - |\downarrow_1\uparrow_2\rangle); \text{ equivalently, } \frac{1}{\sqrt{2}}(|01\rangle - |10\rangle), \qquad (2.2)$$

where, e.g., the term $|01\rangle$ represents the tensor product $|0\rangle \otimes |1\rangle$, and the subscripts 1 and 2 in (2.2) denote, respectively, the spins of particles 1 and 2. The state in Equation (2.2) is an entangled state, i.e., we cannot have a factoring

$$\frac{1}{\sqrt{2}}(|01\rangle - |10\rangle) = x \otimes y,$$

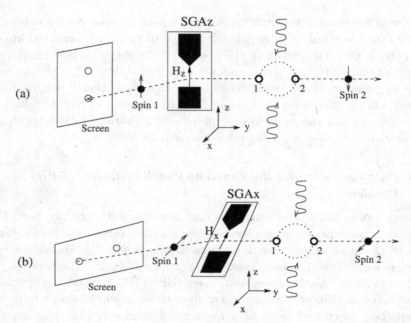

Fig. 4. A dimer, such as H_2, is disintegrated. One particle moves to the left, and the other to the right. We align an SGA along the z-axis and the x-axis, respectively, in (a) and (b). In (a), if the left particle is detected to be spin-up on the screen, then the right particle must be spin-down, $|\downarrow\rangle$. In (b), if the left particle is detected in spin-right, then the right particle must be spin-left, $|\leftarrow\rangle$.

for any single-spin states x and y. Equation (2.2) implies that if we measure the spin of particle 1 to be ↑ (equivalently, 0), then that of particle 2 will be ↓ (equivalently, 1), and vice versa.

At time $t = 0$, we disintegrate the dimer but do not disturb the spins in any way. Then the two broken-up parts move off to opposite directions. See two experimental arrangements in Figure 4(a) and (b).

From the experiments depicted in Figure 4, EPR argue that: "Since at the time of measurement the two systems no longer interact, no real change can take place in the second system in consequence of any thing that may be done to the first system. This is, of course, merely a statement of what is meant by the absence of any interaction between the two systems."

To give a concrete mathematical resolution of this paradox, Bell proposed a verifiable inequality that any theory which satisfies *realism* and *locality* should satisfy. Specifically consider the following thought experiment. C prepares two particles, one of which he sends to A (Alice) and the other to B (Bob). A, upon receipt of her particle, performs one of two measurements randomly on her particle. Call these measurements Q and R (each performed with probability $\frac{1}{2}$). Similarly B performs one of two measurements S or T, each with probability $\frac{1}{2}$ on his particle. All these four measurements take only the values $+1$ or -1. We make the following two assumptions about the experiment:

Realism: The physical properties which the above measurements seek to explore exist independent of observation.

Locality: A's measurements on her particle has no influence on the outcome of B's measurements on his particle and vice-versa.

We now give a rather concise discussion. A somewhat more detailed description may be found in [39], and the original technical source is [12]. Under these assumptions it can be shown that

$$E(QS + RS + RT - QT) \leq 2,$$

where E is the probability expectation. The above inequality is called both the CHSH inequality (named after the four authors of [12] and (one of) Bell's inequalities. Now consider a quantum system of two qubits prepared in the state

$$|\psi\rangle = \frac{1}{\sqrt{2}}(|01\rangle - |10\rangle).$$

Let the first qubit be sent to A and the second to B. Let their corresponding measurements be $Q = \sigma_z \otimes I_2$, $R = \sigma_x \otimes I_2$, $S = -\frac{1}{\sqrt{2}}I_2 \otimes (\sigma_z + \sigma_x)$, $T = \frac{1}{\sqrt{2}}I_2 \otimes (\sigma_z - \sigma_x)$, where σ_x and σ_z are the standard Pauli matrices (see Subsection 3.2) and I_2 is the 2×2 identity matrix.

Then a straightforward calculation shows that

$$E(QS + RS + RT - QT) = 2\sqrt{2} > 2.$$

Thus this system violates the CHSH inequality. Thus, at least one of realism and locality must be violated. Subsequent experimental evidence has vindicated the actual violation of Bell type inequalities.

There is a lesson drawn from the EPR problem: Statements such as "particle 2 is in the spin-down state $| \downarrow_2 \rangle$" already *before* particle 1 is found in the spin-up state $| \uparrow_1 \rangle$ are meaningless. The property of being a pure-state spin-up particle, say, is not possessed by particle 2 before an up/down measurement on particle 1 has found it spin-down. This is quite analogous to the situation discussed in the *quantum eraser* ([32, 33]).

The EPR paper suggested that quantum mechanics was *incomplete* or only a kind of partial theory. That is, just as thermodynamics is a gross (macroscopic) theory of the underlying statistical mechanics, perhaps there are deeper hidden variables for which quantum mechanics plays the role of a kind of "macro-cover theory" (in the spirit of thermodynamics). In particular, Einstein hoped that quantum theory could be supplemented by some additional "hidden" variables.

As another consequence, we have deduced that there are nonlocal correlations in the real world. Hence there are no local hidden variable theories.

Let us argue a little bit more about "locality" and "causality" from the EPR lesson. Consider the classical correlations of two balls in colors of white and black, respectively. You take one at random, but do not look at it and travel to the Moon, and leave the other one on Earth. The very moment you look at the ball on the Moon, you know for sure that the other on Earth has the opposite color! This does not imply that information is flowing instantaneously from the Moon to the Earth, of course. The tricky thing about spin-1/2 is that this argument holds for *any* orientation of the SGA's. Correlated measurements along up/down could equivalently be described by the above classical correlations.

Another conclusion one draws from the above study which leads to the violation of the CHSH inequality is that states such as these properties which could conceivably be harnessed for quantum information processing tasks. This is indeed correct. Such states (together with cleverly chosen measurements) enable tasks such as teleportation, dense coding, cryptography (though the original quantum cryptography protocol BB84 does not require such entangled states).

2.3. *Reversibility and irreversibility, the Landauer principle*

Calculations in electronic digital computers are done by logic gating operations based on the binary Boolean logic. The basic Boolean logic consists of three Boolean operations: $\neg$ (NOT), $\wedge$ (AND), $\vee$ (OR), as indicated by the following tables.

Actually, we only need $\neg$ and $\wedge$ two operations because

$$x \vee y = \neg[(\neg x) \wedge (\neg y)],$$

and thus, the "OR" operation $\vee$ is expressible in terms of $\neg$ and $\wedge$. Similarly, the "AND" operation $\wedge$ is expressible in terms of $\neg$ and $\vee$.

There are other operations such as "COPY," defined as

$$\text{COPY } x = xx,$$

Table 1. Truth tables for ¬ (NOT), ∧ (AND) and ∨ (OR). Note that ¬ is reversible, but ∧ and ∨ are not reversible.

	x	$\neg x$		x	y	$x \wedge y$		x	y	$x \vee y$
				0	0	0		0	0	0
(a)	0	1	(b)	0	1	0	(c)	0	1	1
	1	0		1	0	0		1	0	1
				1	1	1		1	1	1

"FAN-OUT"

$$x \longrightarrow \boxed{FO} \Longleftarrow \begin{matrix} x \\ x \\ x \\ x \end{matrix} \quad \text{(multi-copies of the input } x \text{)}$$

among others.

Now, we turn to a seemingly totally different topic of computer *heating*. The design and operation of the modern electronic computer engenders serious heating of the machine which is a major problem. Such heat dissipation often causes malfunctioning of the machine and is an obstacle to computer circuits miniaturization; it has to be countered by expensive and sophisticated cooling devices and facilities. During the early 1960s, researchers at IBM, led by Rolf Landauer, took on this problem in the hope of designing computers that are energy efficient and not plagued by overheating.

Historically, many scientists (including Brillouin and von Neumann) believed that there is an intrinsic cost involved in information processing, such as the execution of a logic operation by CPU, the copying of information from one memory medium to another, and the measurements of outputs from information flow. For example, there is a so-called Szilard's Principle, which states that it is information gathering that requires an increase in entropy. The finding of the IBM researchers was astounding: that belief was actually a *misconception*. There is no intrinsic irreducible thermodynamic cost required of information processing, acquisition and measurement. However, the *logical reversibility* or *irreversibility* such as shown in Table 1 determines whether there is a minimum cost associated with the information processing. This is *Landauer's principle*, often regarded as the basic principle of the thermodynamics of information processing. It will be stated more precisely in the following paragraphs.

To explain why this is true will take too much space. Here let us just try to offer some insights. We first recall some history of thermodynamics, information theory and the second law of thermodynamics.

The second law of thermodynamics, first formulated by Rudolph Clausius in 1850, states that the entropy of any totally isolated system not at a thermal equilibrium will tend to increase with respect to time, approaching a maximum value. Here, the entropy S is defined through its increment dS by

$$dS = \frac{dQ}{T}, \quad \text{satisfying} \quad dS \geq 0,$$

where dQ is the transfer in energy in the form of heat, and T is the temperature.

For macroscopic systems, the statistical formulation of entropy implies that the second law is overwhelmingly likely to be accurate. However, a non-zero probability is allowed for it to be measurably inaccurate. The great physicist J. C. Maxwell, in his study of thermodynamics, posed an intelligent being, later referred to as *Maxwell's Demon*, to demonstrate that there could be situations in which the second law is violated. The violation of the second law of thermodynamics is a big controversy and Maxwell's Demon attracted the attention of many scientists. Szilard [37] (1929) suggested a simplified version. An excellent discussion may be found schematically in Plenio and Vitelli [29].

In Maxwell's supposition, an element of *consciousness* or *intelligence* entered into the realm of physical science, as the Demon were able to sort out molecules with higher and lower energies. The role of such intelligence of the Demon has generated significant debates as to how and whether one ought to characterize it physically. This issue of vagueness involving intelligence can be avoided by the substitution of an automatic mechanism which can perform Demon's functions. (Indeed, an SGA could perform such a function in principle.) With this, the paradox of Maxwell's Demon was still not easy to resolve. A satisfactory exorcism of the Demon was finally made by Charles H. Bennett in [3] (1982) by applying Landauer's principle. In Landauer's study [25] (1961), he had an important insight about a fundamental asymmetry in information processing. For example, the negation of a binary variable or the copying of classical information, cf. Table 1(a), can be done reversibly without expending any energy. But when information is erased, or if an irreversible binary logic operation such as $a \wedge b = c$ is executed, where $\wedge$ is the "AND" operation as introduced in Table 1(b), there is always an energy cost of $kT \ln 2$ to be consumed, where k is the Boltzmann constant and T is the absolute temperature. In information theory, consider a system with n events (or states), each with probability of occurrence p_i, for $i = 1, 2, \ldots, n$, where $0 < p_i \leq 1$ and $p_1 + p_2 + \cdots + p_n = 1$. Then the total information "value" of the system is (defined by) the Shannon entropy

$$H_S = -\sum_{i=1}^{n} p_i \log p_i, \quad (\log = \text{logarithm with base 2}). \qquad (2.3)$$

The above entropy can be linked to the Boltzmann entropy

$$H_B = -k \ln 2 \sum_{i=1}^{n} p_i \log p_i, \quad (\ln = \text{the natural logarithm}) \qquad (2.4)$$

by considering (from statistical mechanics) how many arrangements there are in the assemblage of matter and energy in a physical system. The term $k \ln 2$ is a scaling factor in the conversion from the classical Shannon information entropy (2.3) to the Boltzmann entropy (2.4). For a (1-bit) binary system, the single 1-bit 0 and 1 occur with equal probability $p_1 = p_2 = 1/2$, so we obtain

$$H_B = -k \ln 2 \left[\frac{1}{2} \log \left(\frac{1}{2} \right) + \frac{1}{2} \log \left(\frac{1}{2} \right) \right] = k \ln 2. \qquad (2.5)$$

Landauer argues that in any irreversible computing where information is lost, forgotten or erased, the entropy so obtained is the amount of thermodynamical entropy you will generate in erasing the information. The least of which is one bit of information. Thus, the Boltzmann entropy is (2.5), which after multiplying by T, gives the minimum of expended energy $kT \ln 2$ (dissipated into the environment). This is the Landauer principle of information erasure. From this principle, one can actually derive the second law of thermodynamics [29].

The discussions in this section so far are all classical. But now let us be equipped with quantum mechanics and revisit Maxwell's Demon. What Maxwell envisioned in his Demon's function having the ability to sort out molecules with faster and slower speeds is no longer as impossible or far-fetched as Maxwell originally thought. Indeed, an SGA can certainly achieve that in principle, where higher energy and lower energy atoms in an atom beam are separated (i.e., sorted out) by an SGA. The tracking and detection of their trajectories can be done, e.g., by cavity-QED (see Subsection 3.3, e.g.) and photo-detectors. However, in the quantum realm, some unexpected, amazing phenomena happen. When we observe and determine the path of an atom in a double-slit Young's experiment, the situation is as follows:

(i) If the path information is fully available (irrespective of whether there is any human knowledge about it), then there are no interference fringes;

(ii) If we have interference fringes of perfect visibility, then there cannot be any path knowledge available.

However, it does not follow that whenever there are no fringes, then path knowledge must be available. Rather, it is quite easy to have a situation where neither path knowledge nor fringes are available. Even then, it may be possible to restore fringes by a "postselection" procedure called "quantum erasure". The nomenclature of "quantum erasure" — like so many other terms — could be somewhat misleading. Key studies of the quantum eraser process are [32, 33], and experimentally confirmed in [23, 41]. This puzzle of quantum eraser turns out to be a quantum mechanical analogue of Maxwell's Demon. Since information erasure has taken place in this process, Landauer's principle applies to the quantum eraser.

In order to discuss the concept and setup of the quantum eraser, some prerequisite on cavity-QED is necessary. We thus refer the discussions to [10] as well as Subsection 3.3.

An elementary introduction to Maxwell's Demon, Szilard's engine and quantum eraser can be found in Scully and Scully [35].

As a final note of this section, we mention that all quantum processes are unitary and reversible. This follows from the fact that the evolution of a quantum system follows the Schrödinger equation

$$i\hbar\frac{\partial}{\partial t}\psi(\vec{x},t) = H(\vec{x},t)\psi(\vec{x},t),$$

where there is a 2-parameter family of evolution operator U(s,t) which in some way constitutes a unitary group. Therefore, the logic operations in quantum computing, based on unitary operations, are always reversible (in "idealistic environments"). This is a major distinction between classical computing and quantum computing. Thus, ideally, a quantum computer can carry out computations without any energy cost and problem associated with heating. Nevertheless, we need to keep in mind that in reality, the environment is mostly not idealistic. If the situation is such that there is strong coupling between the system and reservoir, then information and energy loss occurs in between in a rather non negligible way.

Quantum computing is a fascinating subject. Several schemes and devices have shown promise, but each has its stumbling roadblocks and severe challenges. In turn, each has generated cutting-edge technologies that spin off into computing and other areas useful for the economy and defense.

The latest assessments on quantum computing may be found in an article in Nature by P. Ball [1]. A poignant point made therein is the quote from A. Steane of the Oxford Computing Group: "A useful (quantum) computer by 2020 is realistic". That article also quoted S. Lloyd of MIT as saying that fully functioning quantum simulators will be readily available by 2020.

These represent a dramatic increase of confidence over five years ago, when most of the researchers if asked how long it would take to make a genuine quantum computing machine, they probably would have said it was too far off even to guess. D. Deutsch of the Oxford University optimistically added that "a practical quantum computer may well be achieved within the next decade".

3. Quantum Computing Devices

3.1. *The electromagnetic field and its quantization*

There are at least a dozen major proposals for implementing quantum computing on quantum devices; see [10], also [26]. The great majority of such computing operations is achieved through the application of *electromagnetic* or *laser* fields on atoms and molecules. Such fields, classically, can be represented as follows: the electrical field vector E and the magnetic field H in a sourceless region are coupled in the Maxwell equations:

$$\nabla \times H = \frac{\partial D}{\partial t}$$
$$\nabla \times E = -\frac{\partial B}{\partial t}$$
$$\nabla \cdot B = 0 \tag{3.1}$$
$$\nabla \cdot D = 0$$

where $D = \epsilon_0 E$, $B = \mu_0 H$, and ϵ_0 and μ_0 are the free space permittivity and permeability, respectively. It is also known that $\frac{1}{\epsilon_0 \mu_0} = c^2$ where c is the speed of light in vacuum.

Consider an optical field in a rectangle cavity of length L and volume V. For the purpose of simplification, we assume the electrical field is linearly polarized in the x-direction and vanishes at the two end surface. Along the direction of z-axis, the electric field is a superposition of sinusoidal functions with different frequencies,

$$E(z,t) = \sum_j A_j q_j(t) \sin(k_j z) e_x, \tag{3.2}$$

where each oscillating $\sin(k_j z)e_x$ is called a *normal mode* with $k_j = j\pi/L$, $j = 1, 2, 3, \ldots$, and $q_j(t)$ is the amplitude of mode j. The coefficient A_j is defined as

$$A_j = \left(\frac{2\nu_j^2 m_j}{V\epsilon_0} \right)^{1/2}, \tag{3.3}$$

and $\nu_j = ck_j = j\pi c/L$ is the frequency corresponding to the j-th normal mode. The constant m_j has the unit of mass, and it is deliberately introduced to establish the homologues between the Hamiltonian of a single mode electrical field and the Hamiltonian of a mechanical harmonic oscillator.

The magnetic field can be calculated using Maxwell's equation (3.1) as

$$\boldsymbol{H} = e_y \epsilon_0 \sum_j \frac{A_j \dot{q}_j(t)}{k_j} \cos(k_j z). \tag{3.4}$$

Together with Equations (3.2) and (3.4), we have the electromagnetic field in the cavity. The classical Hamiltonian then can be obtained as an integral over the volume V,

$$\begin{aligned} H &= \frac{1}{2} \int_V (\epsilon_0 \|\boldsymbol{E}\|^2 + \mu_0 \|\boldsymbol{H}\|^2) dV \\ &= \sum_j \frac{1}{2} \nu_j^2 m_j q_j(t)^2 + \frac{1}{2} m_j \dot{q}_j(t)^2. \end{aligned} \tag{3.5}$$

The two terms after the summation sign actually can be regarded as the potential and kinetic energies of a harmonic oscillator. The analogy between the Hamiltonian of a cavity and that of a set of independent mechanical harmonic oscillators implies that we can quantize the cavity field in a way similar to how we quantize a harmonic oscillator [34].

Define $p_j = m_j \dot{q}_j$ as the canonical momentum of the j-th mode, then we can rewrite H as

$$H = \frac{1}{2} \sum_j \left(m_j \nu_j^2 q_j^2 + \frac{p_j^2}{m_j} \right). \tag{3.6}$$

To quantize the field, we treat q_j and p_j as operators, and as in quantizing a harmonic oscillator, they follow the commutation relations

$$[q_j, p_{j'}] = i\hbar \delta_{jj'}, \ [q_j, q_{j'}] = [p_j, p_{j'}] = 0. \tag{3.7}$$

Two other operators a and $a^\dagger$ are defined as

$$a_j^\dagger e^{i\nu_j t} = \frac{1}{\sqrt{2m\hbar\nu}}(m_j\nu_j q_j - ip_j),$$
$$a_j e^{-i\nu_j t} = \frac{1}{\sqrt{2m\hbar\nu}}(m_j\nu_j q_j + ip_j).$$

(3.8)

From (3.7), it is easy to show that

$$[a_j, a_{j'}^\dagger] = \delta_{jj'}, \ [a_j, a_{j'}] = [a_j^\dagger, a_{j'}^\dagger] = 0.$$

(3.9)

Since different modes are independent, we can focus on a single mode and neglect the rest. Simple calculation shows that the Hamiltonian of a single mode field in terms of a and $a^\dagger$ is given by

$$H = \frac{1}{2}\left(m\nu^2 q^2 + \frac{p^2}{m}\right)$$
$$= \hbar\nu\left(a^\dagger a + \frac{1}{2}\right).$$

(3.10)

The term $1/2$ in the last expression follows from the commutation relations of a and $a^\dagger$.

Let $|n\rangle$ be the eigenfunction of H with eigenvalue E_n. Then it can be shown that $a|n\rangle$ is also an eigenfunction of H:

$$Ha|n\rangle = \hbar\nu\left(a^\dagger a + \frac{1}{2}\right)a|n\rangle$$
$$= a(E_n - \hbar\nu)|n\rangle$$
$$= (E_n - \hbar\nu)a|n\rangle.$$

(3.11)

Denote $|n-1\rangle$ as the normalized eigenfunction from $a|n\rangle$ we can get a series of eigenfunction of H using a and $a^\dagger$ as $|n\rangle$, $n = 0, 1, 2, \ldots$. Furthermore, we have

$$a|n\rangle = \sqrt{n}|n-1\rangle \quad \text{and} \quad a^\dagger|n\rangle = \sqrt{n+1}|n+1\rangle.$$

(3.12)

Thus $a^\dagger$ is called the creation operator, and a is called the annihilation operator. The above leads to

$$a^\dagger a|n\rangle = a^\dagger \sqrt{n}|n-1\rangle = n|n\rangle,$$

(3.13)

and

$$H|n\rangle = \left(n + \frac{1}{2}\right)\hbar\nu.$$

(3.14)

In terms of $a^\dagger$ and a, the quantized single mode electric field can be written as

$$E(z,t) = e_x \left(\frac{\hbar\nu}{\epsilon_0 V}\right)^{1/2} (ae^{-i\nu t} + a^\dagger e^{i\nu t})\sin(kz). \qquad (3.15)$$

3.2. The interaction between a two-level atom and the optical field

An atom has infinitely many energy levels. If there exist certain two energy levels such that their energy difference is equal or very close to the energy of a photon, $\hbar\nu$, where ν is the frequency of the optical filed, the optical field is resonant. The atom absorbs and emits photons while it jumps between the two energy levels. Since the energy levels of an atom are not evenly distributed, other levels are detuned from the resonance and we can treat the atom as a two-level atom. Optical fields can be treated classically or quantized depending on its intensity. In this subsection, we first take it as a classical field and introduce the Rabi rotation, leading to a cavity-QED.

We consider atoms such as hydrogen or the alkali family. The alkali atoms only have one electron on the outmost shell and they behave similar to hydrogen atoms. The Hamiltonian of such an atom in a single mode optical field has two parts, H_0 and H_{al}, which are called the unperturbed Hamiltonian and the interaction Hamiltonian, respectively. The total Hamiltonian reads

$$H = H_0 + H_{al}. \qquad (3.16)$$

In general, the size of the atom is far smaller or shorter than the wavelength of the optical field. Thus, we can invoke the electric dipole approximation and obtain [34]

$$H_{al} = -e\boldsymbol{r} \cdot \boldsymbol{E}(\boldsymbol{r}_0, t), \qquad (3.17)$$

where $\boldsymbol{r}_0$ is the position of the atom, and $\boldsymbol{r}$ is the position operator of the electron. When the electric field is polarized along the x-axis and propagating along the z-axis, this is simplified to

$$H_{al} = -exE(z,t). \qquad (3.18)$$

Denote the two energy levels involved as $|e\rangle$ and $|g\rangle$, where $|e\rangle$ is the excited level with eigenvalue E_e and $|g\rangle$ is the ground level with eigenvalue

E_g. Then the unperturbed Hamiltonian can be written as

$$\begin{aligned}
H_0 &= E_g|g\rangle\langle g| + E_e|e\rangle\langle e| \\
&= \frac{E_g + E_e}{2}(|g\rangle\langle g| + E_e|e\rangle\langle e|) + \frac{\hbar\omega}{2}(|e\rangle\langle e| - |g\rangle\langle g|) \\
&= \frac{E_g + E_e}{2}I + \frac{\hbar\omega}{2}(|e\rangle\langle e| - |g\rangle\langle g|),
\end{aligned} \tag{3.19}$$

where $\hbar\omega = E_e - E_g$. Since a multiple of the identity matrix only contributes a uniform phase shift for both states, it can be dropped and we obtain

$$H_0 = \frac{\hbar\omega}{2}(|e\rangle\langle e| - |g\rangle\langle g|). \tag{3.20}$$

Similarly, we obtain

$$\begin{aligned}
H_{al} &= -exE(z,t) \\
&= -e(|g\rangle\langle g| + |e\rangle\langle e|)x(|g\rangle\langle g| + |e\rangle\langle e|)E(z,t).
\end{aligned} \tag{3.21}$$

The symmetry of the wavefunction implies that both $\langle g|x|g\rangle$ and $\langle e|x|e\rangle$ vanish. The interaction Hamiltonian H_{al} becomes

$$H_{al} = -(\rho|e\rangle\langle g| + \rho|g\rangle\langle e|)E(z,t), \tag{3.22}$$

where $\rho = \langle e|x|g\rangle$, and we assume that ρ is real.

Since the atom lives in the space spanned by $|g\rangle$ and $|e\rangle$, its wavefunction at time t, denoted as $|\psi(t)\rangle$, can be written as

$$|\psi(t)\rangle = c(t)|e\rangle + b(t)|g\rangle. \tag{3.23}$$

The Schrödinger equation gives

$$\begin{aligned}
\frac{d}{dt}|\psi(t)\rangle &= \dot{c}(t)|e\rangle + \dot{b}(t)|g\rangle \\
&= -\frac{i}{\hbar}H|\psi(t)\rangle = -\frac{i}{\hbar}(H_0 + H_{al})|\psi(t)\rangle.
\end{aligned} \tag{3.24}$$

With respect to the ordered basis $\{|g\rangle, |e\rangle\}$, the matrix form of the total Hamiltonian H is

$$H = \begin{bmatrix} -\dfrac{\hbar\omega}{2} & -\rho E(z,t) \\ -\rho E(z,t) & \dfrac{\hbar\omega}{2} \end{bmatrix}. \tag{3.25}$$

Similarly, we have H_{al} as

$$H_{al} = \begin{bmatrix} 0 & -\rho E(z,t) \\ -\rho E(\dot{z},t) & 0 \end{bmatrix}. \tag{3.26}$$

We define $|\phi(t)\rangle = U_0|\psi(t)\rangle = ce^{i\omega t/2}|e\rangle + be^{-i\omega t/2}|g\rangle$, where

$$U_0 = e^{\frac{i\,H_0\,t}{\hbar}}$$
$$= \begin{bmatrix} e^{-i\omega t/2} & 0 \\ 0 & e^{i\omega t/2} \end{bmatrix}. \tag{3.27}$$

In other words, we put the system in a "rotating frame". By substituting it back to (3.24), we obtain the Schrödinger equation for $|\phi\rangle$:

$$\frac{\partial}{\partial t}|\phi(t)\rangle = -\frac{i}{\hbar}\mathcal{H}|\phi(t)\rangle, \tag{3.28}$$

where $\mathcal{H}$ is called the Hamiltonian in the *interaction picture* [34], obtained via

$$\mathcal{H} = U_0 H_{al} U_0^\dagger. \tag{3.29}$$

We then have

$$\mathcal{H} = \begin{bmatrix} e^{-i\omega t/2} & 0 \\ 0 & e^{i\omega t/2} \end{bmatrix} \begin{bmatrix} 0 & -\rho E(z,t) \\ -\rho E(z,t) & 0 \end{bmatrix} \begin{bmatrix} e^{i\omega t/2} & 0 \\ 0 & e^{-i\omega t/2} \end{bmatrix}$$
$$= \begin{bmatrix} 0 & -\rho E(z,t)e^{-i\omega t} \\ -\rho E(z,t)e^{i\omega t} & 0 \end{bmatrix}. \tag{3.30}$$

We assume the light field is resonant at frequency ω with phase α:

$$E(z,t) = \epsilon\cos(\omega t + \alpha) = \frac{\epsilon}{2}(e^{(\omega t + \alpha)i} + e^{-(\omega t + \alpha)i}). \tag{3.31}$$

Then $\mathcal{H}$ can be further simplified to

$$\mathcal{H} = \hbar\Omega \begin{bmatrix} 0 & e^{i\alpha} \\ e^{-i\alpha} & 0 \end{bmatrix} = \hbar\Omega(\cos(\alpha)\sigma_x - \sin(\alpha)\sigma_y), \tag{3.32}$$

where $\Omega = \frac{-\rho\epsilon}{2\hbar}$, called the Rabi frequency, and σ_x and σ_y are Pauli matrices.[a] Note that high frequency terms such as $e^{2i\omega t}$ and $e^{-2i\omega t}$ are dropped since their effect can not be observed in experiments.

After time duration t, the new $|\phi(t)\rangle$ is found as

$$|\phi(t)\rangle = e^{-i\Omega(\sigma_x\cos(\alpha) - \sigma_y\sin(\alpha))t}|\phi(0)\rangle. \tag{3.33}$$

---

[a]The Pauli matrices are $\sigma_x = \begin{bmatrix} 0 & 1 \\ 1 & 0 \end{bmatrix}$, $\sigma_y = \begin{bmatrix} 0 & -i \\ i & 0 \end{bmatrix}$, and $\sigma_z = \begin{bmatrix} 1 & 0 \\ 0 & -1 \end{bmatrix}$.

Further calculation shows that the electron jumps between $|g\rangle$ and $|e\rangle$ with frequency Ω. The evolution operator is obtained as

$$U_{\theta/2,\alpha} = e^{-i\Omega(\sigma_x \cos(\alpha) - \sigma_y \sin(\alpha))t}$$

$$= \begin{bmatrix} \cos\left(\dfrac{\theta}{2}\right) & -i\sin\left(\dfrac{\theta}{2}\right)e^{i\alpha} \\ -i\sin\left(\dfrac{\theta}{2}\right)e^{-i\alpha} & \cos\left(\dfrac{\theta}{2}\right) \end{bmatrix}, \tag{3.34}$$

where $\theta = 2\Omega t$. This is a *1-bit rotation operator*, also called a *Rabi rotation gate*.

3.3. *Cavity-QED*

An optical cavity has a set of highly reflective mirrors to keep photons within for a long time, see Figure 5. When an atom passes through the cavity, it interacts with the cavity field and gets entangled with that field. Atoms could be injected into the cavity via an optical lattice. The interaction time is determined by how long the atom stays in the cavity. Atoms have long coherence time, and photons are known for their use in long distance communication. Thus cavity-QED has useful potential for quantum information processing or quantum computation.

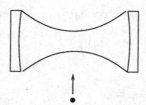

Fig. 5.   A schematic of an optical cavity. Highly reflective mirrors keep photons in the cavity for a long time. An atom is injected into that cavity, leading to interactions between the atom and photons.

We focus on a single mode cavity and an atom within. The cavity field is close to the resonant frequency between two levels of the atom, denoted as $|\alpha\rangle$ and $|\beta\rangle$, respectively. From the calculations of the preceding section, we see that the total Hamiltonian of the atom-cavity system includes three parts [10]:

$$H = H_0 + H_1 + H_2, \tag{3.35}$$

with

$$H_0 = \frac{\hbar\omega}{2}(|\alpha\rangle\langle\alpha| - |\beta\rangle\langle\beta|),$$

$$H_1 = \hbar\nu a^\dagger a, \tag{3.36}$$

$$H_2 = \hbar g(|\alpha\rangle\langle\beta|a + |\beta\rangle\langle\alpha|a^\dagger) \tag{3.37}$$

where the interaction constant is $g = |\rho_{\alpha\beta}| \left(\frac{\nu}{\hbar V \epsilon_0}\right)^{1/2}$, and $\rho_{\alpha\beta}$ is the electric dipole transition moment. Here we treat the cavity field as quantized. The operators $a^\dagger$ and a are defined as before. Constant terms in the Hamiltonian, such as the $1/2$ in the Hamiltonian of laser field, can be simply ignored. Let $|n\rangle$ be the number state of the cavity field. The subspace spanned by $\{|\alpha, n-1\rangle, |\beta, n\rangle\}$ is closed under the operator H. The matrix form of H with respect to the ordered basis $\{|\alpha, n-1\rangle, |\beta, n\rangle\}$ is

$$H = \hbar \begin{bmatrix} \frac{1}{2}\omega + \nu(n-1) & g\sqrt{n} \\ g\sqrt{n} & -\frac{1}{2}\omega + \nu n \end{bmatrix}. \tag{3.38}$$

Theorem 3.1. *The Hamiltonian given in (3.38) has two eigenfunctions*

$|+\rangle_n = \cos\theta_n|\alpha, n-1\rangle - \sin\theta_n|\beta, n\rangle$ *with eigenvalue*

$$E_+^n = \hbar\left(n\nu + \frac{1}{2}(-\nu - \Omega_n)\right)$$

$|-\rangle_n = \sin\theta_n|\alpha, n-1\rangle + \cos\theta_n|\beta, n\rangle$ *with eigenvalue* $\qquad$ (3.39)

$$E_-^n = \hbar\left(n\nu + \frac{1}{2}(-\nu + \Omega_n)\right)$$

where

$$\Delta = \nu - \omega,$$

$$\Omega_n = (\Delta^2 + 4g^2 n)^{1/2},$$

$$D = ((\Omega_n - \Delta)^2 + 4g^2 n)^{1/2}, \tag{3.40}$$

$$\sin\theta_n = \frac{\Omega_n - \Delta}{D}, \quad \cos\theta_n = \frac{2g\sqrt{n}}{D}.$$

In the limit when $|\Delta| \gg 2g\sqrt{n}$, called the large detuning limit, the off diagonal elements of H approach zero. We have

$$|+\rangle_n \approx |\alpha, n-1\rangle, E_+^n = \hbar\left(\frac{\omega}{2} + \nu(n-1)\right) - \frac{\hbar g^2 n}{\Delta},$$

$$|-\rangle_n \approx |\beta, n\rangle, E_-^n = \hbar\left(-\frac{\omega}{2} + \nu n\right) + \frac{\hbar g^2 n}{\Delta}. \tag{3.41}$$

This leads to a diagonalized effective Hamiltonian for H_2 given by

$$\bar{H}_2 = -\frac{\hbar g^2}{\Delta}(|\alpha\rangle\langle\alpha|aa^\dagger - |\beta\rangle\langle\beta|a^\dagger a). \tag{3.42}$$

Let $|\gamma\rangle$ be another energy level of the atom and it is far detuned from the cavity. We decode the logic state of the atom using $|\beta\rangle$ and $|\gamma\rangle$, by designating $|\beta\rangle$ as $|1\rangle$ and $|\gamma\rangle$ as $|0\rangle$. Assume there is at most one photon in the cavity, thus $n = 1$ and the space has a basis of $\{|00\rangle, |01\rangle, |10\rangle, |11\rangle\}$ with

$$|00\rangle = |\gamma, 0\rangle, \ |01\rangle = |\gamma, 1\rangle$$
$$|10\rangle = |\beta, 0\rangle, \ |11\rangle = |\beta, 1\rangle. \tag{3.43}$$

Theorem 3.2. *With respect to the ordered basis* $\{|00\rangle, |01\rangle, |10\rangle, |11\rangle\}$, *the effective Hamiltonian* $\bar{H}_2$ *is diagonal and*

$$\bar{H}_2 = \begin{bmatrix} 0 & & & \\ & 0 & & \\ & & 0 & \\ & & & \dfrac{\hbar g^2}{\Delta} \end{bmatrix}. \tag{3.44}$$

The Hamiltonian in (3.44) can be used to form a conditional phase gate.

3.4. Optical lattices

Optical lattices are essential in the design of quantum computation using cold atoms because of its capability to trap and transport atoms in space. An optical lattice could be one, two, or three dimensional. A basic configuration is a one dimensional optical lattice formed by two counter-propagating light beams which are linearly polarized and have the same amplitude. Depending on the angle between the polarization directions of the two light beams, the polarization of the light field changes along the propagation axis.

3.4.1. Setting up the fields

Without loss of generality, we assume that both light beams are traveling along the z-axis and their electric fields are given by

$$E_1(z, t) = \frac{1}{2}\epsilon e_1 e^{-i(\omega t - kz)} + c.c.,$$

$$E_2(z, t) = \frac{1}{2}\epsilon e_2 e^{-i(\omega t + kz)} + c.c., \tag{3.45}$$

Fig. 6.   The schematic of a linθlin configuration. Two linearly polarized light beams travel along the z-axis in opposite directions, with the angle between the polarizations θ.

respectively, where *c.c.* means complex conjugate. Let the angle between the two polarization vectors e_1 and e_2 be θ. We arrange the x and y axes so that

$$
\begin{aligned}
e_1 &= e_x \cos\left(\frac{\theta}{2}\right) + e_y \sin\left(\frac{\theta}{2}\right), \\
e_2 &= e_x \cos\left(\frac{\theta}{2}\right) - e_y \sin\left(\frac{\theta}{2}\right).
\end{aligned}
\tag{3.46}
$$

This configuration, as shown in Figure 6, is called a linθlin *configuration*. The total field along the z-axis is then obtained as

$$
\begin{aligned}
E(z,t) &= E_1 + E_2 \\
&= \frac{1}{2}\epsilon e^{-i\omega t}\left(2e_x \cos\frac{\theta}{2}\cos(kz) + 2ie_y \sin\frac{\theta}{2}\sin(kz)\right) \\
&\quad + c.c.
\end{aligned}
\tag{3.47}
$$

We are mostly interested in the distribution of circular polarizations. It is convenient to use another basis of e_+ and e_-, defined by

$$
\begin{aligned}
e_+ &= -\frac{1}{\sqrt{2}}(e_x + ie_y), \\
e_- &= \frac{1}{\sqrt{2}}(e_x - ie_y),
\end{aligned}
\tag{3.48}
$$

corresponding to the left circular polarization and right circular polarization, respectively. In terms of e_+ and e_-, the electric field can be rewritten as

$$E(z,t) = \frac{\sqrt{2}}{2}\epsilon e^{-i\omega t}\left(-e_+ \cos\left(kz - \frac{\theta}{2}\right) + e_- \cos\left(kz + \frac{\theta}{2}\right)\right)$$

$$+ c.c.$$

$$= \frac{1}{2}(A_+ e_+ + A_- e_-)e^{-i\omega t} + c.c., \tag{3.49}$$

where $A_+ = -\sqrt{2}\epsilon\cos(kz - \frac{\theta}{2})$ and $A_- = \sqrt{2}\epsilon\cos(kz + \frac{\theta}{2})$, which are the amplitude of the left and right circularly polarized parts, respectively.

Changing the angle θ causes changes of the polarization along the z-axis. When $\theta = 0$, $A_- = -A_+$, the light field is linearly polarized along the x-axis everywhere. When θ is different from zero, the polarization shows a gradient along the propagation direction. The highest intensity of the left circularly polarized part, A_+^2, occurs at $z = \frac{m}{2}\lambda + \frac{\theta}{2k}$, where m is an integer and $\lambda = \frac{2\pi}{k}$, while that of A_- occurs at $z = \frac{m}{2}\lambda - \frac{\theta}{2k}$. The distance between a peak of A_+^2 and its closest neighboring peak of A_-^2 is θ/k. Here we limit θ between 0 and $\pi/2$. Both the A_+ peak and A_- peak move with θ, shown in Figure 7. A special case is $\theta = \pi/2$, when the distance between an A_+ peak and its neighboring A_- peak is largest. It is called a lin $\perp$ lin *configuration* since the polarization vectors are perpendicular to each other. After rearranging the origin or phases, the electric field can be obtained as

$$E(z,t) = \frac{\sqrt{2}}{2}\epsilon e^{-i\omega t}(e_- \cos(kz) - i\, e_+ \sin(kz)) + c.c. \tag{3.50}$$

Along the z-axis, its polarization changes in the sequential order of linear polarization along the x-axis, left circular polarization, linear polarization along the y-axis, and right circular polarization, shown in Figure 8.

3.4.2. *Storing, sorting, and transporting ultra cold neutral atoms using optical lattices*

An oscillating optical field can remove the degeneracy of an atom by inducing different energy shifts for its sublevels with different angular momenta. Such energy shifts, called ac-Stark shifts or light shifts, are proportional to the intensity of the light and depends on the polarization of the light. An optical lattice with polarization gradient thus forms an array of periodic potential wells to store and trap ultra cold neutral atoms.

We consider an alkali atom with only two sublevels in the ground state and three sublevels in the excited state. If j_g and j_e are the angular quantum number for the ground state and excited state, respectively, we have

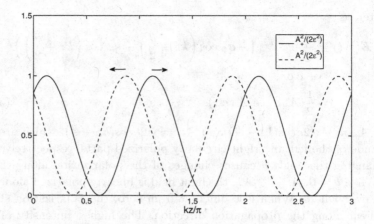

Fig. 7. Illustration of the left and right circular polarization parts along the z-axis. While A_+^2 is plotted in solid line, A_-^2 is plotted in dashed line. The arrow shows the moving direction of the peaks when θ is increasing.

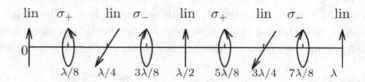

Fig. 8. Changing polarization along the z-axis for a lin $\perp$ lin configuration. At $z = 0$, $\lambda/2$, $\lambda, \ldots$, where $\lambda = 2\pi/k$, the polarization is linear along e_x. At $z = \lambda/4$, $3\lambda/4$, $5\lambda/4, \ldots$, the polarization is linear along e_y. However, at $z = \lambda/8$, $5\lambda/8$, $9\lambda/8, \ldots$, the field has a left circular polarization (σ^+), and at $z = 3\lambda/8$, $7\lambda/8$, $11\lambda/8$, $15\lambda/8, \ldots$, it changes to a right circular polarization (σ^-).

$j_g = 1/2$ and $j_e = 3/2$. The configuration is shown in Figure 9 with the corresponding Clebsch-Gordan coefficients.

According to the *selection rules*, no absorption-emission may cause the transition between $|g_{+1/2}\rangle$ and $|g_{-1/2}\rangle$. We also assume the low saturation limit. The life time of the excited state is so short that the atom actually lives in the space spanned by the two ground sublevels. Thus in the matrix form in terms of $|g_{+1/2}\rangle$ and $|g_{-1/2}\rangle$, the effective Hamiltonian of the atom, H_{eff}, is diagonal, and $|g_{+1/2}\rangle$ and $|g_{-1/2}\rangle$ are eigenfunctions of H_{eff} with eigenvalue $E_{+1/2}$ and $E_{-1/2}$, respectively. By arranging the origin, the electric field in a lin $\perp$ lin configuration can be given as (3.50). Then the

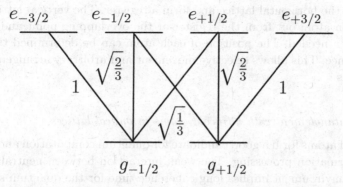

Fig. 9. The configuration of the $J_g = 1/2 \leftrightarrow J_e = 3/2$ transition with the corresponding Clebsch-Gordan coefficients.

potential energy, $E_{+1/2}$ and $E_{-1/2}$, are obtained as [13]

$$E_{+1/2} = -\frac{3U_0}{2} + U_0 \cos^2 kz,$$

$$E_{-1/2} = -\frac{3U_0}{2} + U_0 \sin^2 kz,$$

(3.51)

with

$$U_0 = -\frac{2}{3}\hbar\hat{\delta} = -\frac{2}{3}\hbar\delta s_0,$$

(3.52)

where δ is the detuning frequency and s_0 depends on δ and properties of the atom. Equation (3.51) shows that atoms in state $|g_{+1/2}\rangle$ and those in state $|g_{-1/2}\rangle$ experience different potentials. One is dominated by the left circular polarization part and the other is dominated by the right circular polarization part.

An optical lattice can also be used to transport atoms trapped within. While changing the relative polarization angles between the two light beams causes neighboring atoms move in opposite directions, changing the relative phase between the two beams move the whole array of atoms in one direction. Furthermore, by slightly changing the frequency of one beam, we can transform the static lattice into a "convey belt". Miroshnychenko has used this scheme to sort the atoms in a line [4, 27, 30]. The sorting machine consists of two lattices. One is horizontal and the other intersects with it at a certain point, called the overlapping point. For simplicity, we assume the second lattice is vertical, although it is not necessary the case. Atoms are stored in the horizontal lattice and form a register. The vertical lattice can extract an atom out from the register at the overlapping point and put it back when needed. The position of each atom can be determined through fluorescence. This allows selective movement and arbitrary arrangement of the atoms.

3.4.3. *Entanglement with ultracold atoms in optical lattices*

Ultracold atoms form a good candidate for quantum computation and quantum information processing. The weak interaction between neutral atoms and the environment implies long coherence time for the quantum system. Moreover, the atoms are readily to be cooled, stored, and moved in optical lattices which could hold thousands of atoms at the same time. Although it is hard to entangle arbitrary two atoms in the lattice, neighboring atoms can be coupled via either elastic collision or induced electric dipole-dipole interaction.

Jaksch [22] has suggested to entangle two neutral atoms using elastic collisions. Two ground $S_{1/2}$ hyperfine structures of an alkali atom are used to represent the logic states:

$$
\begin{aligned}
|0\rangle &= |F_\downarrow, m_f = 1\rangle, \\
|1\rangle &= |F_\uparrow, m_f = 2\rangle.
\end{aligned}
\tag{3.53}
$$

It is assumed that the atom has nuclear spin $3/2$. The potential an atom experiences depends on the state of the atom. Potentials of an atom at state $|0\rangle$ and $|1\rangle$ are given by

$$
\begin{aligned}
V^0(z, \theta) &= (V_{m_s=1/2}(z, \theta) + 3V_{m_s=-1/2}(z, \theta))/4, \\
V^1(z, \theta) &= V_{m_s=1/2}(z, \theta),
\end{aligned}
\tag{3.54}
$$

respectively, where $V_{m_s=1/2}$ $(V_{m_s=-1/2})$ is the potential energy experienced by atoms with $m_s = 1/2$ (respectively, $m_s = -1/2$). They are given by

$$
\begin{aligned}
V_{m_s=1/2}(z, \theta) &= \alpha\epsilon^2 \sin^2(kz + \theta/2), \\
V_{m_s=-1/2}(z, \theta) &= \alpha\epsilon^2 \sin^2(kz - \theta/2).
\end{aligned}
\tag{3.55}
$$

As in the previous section, θ is the angle between the polarization directions of the two beams. Common terms are ignored for the sake of simplification.

While changing θ from $\pi/2$ to 0, two neighbor atoms are moved into one well if and only if the left atom is in state $|0\rangle$ and the right atom is state $|1\rangle$ before the change, denoted by $|01\rangle$. Thus the phase change caused by the elastic collision is conditional. Let $|00\rangle$, $|01\rangle$, $|10\rangle$, and $|11\rangle$ be the ordered basis of the wavefunction space. The matrix of the evolution can be written as

$$U = \begin{bmatrix} 1 & & & \\ & e^{-i\phi_{01}} & & \\ & & 1 & \\ & & & 1 \end{bmatrix}, \tag{3.56}$$

where ϕ_{01} is the induced phase change. The operation leads to a controlled phase shift when combined with other 1-bit gates. It is not selective. The same operation is applied to every neighboring pair of atoms in an optical lattice simultaneously.

The weak interaction between two neutral atoms requires long time to complete a 2-bit entanglement. This can be improved by making use of a strong interaction mechanism, such as the induced electric dipole-dipole interaction. In the scheme proposed by Brennen [5, 6], logic states are encoded with the ground $S_{1/2}$ hyperfine structures of an alkali atom with nuclear spin 3/2 as

$$\begin{aligned} |1\rangle_\pm &= |F_\uparrow, M_F = \pm 1\rangle, \\ |0\rangle_\pm &= |F_\downarrow, M_F = \mp 1\rangle, \end{aligned} \tag{3.57}$$

where $+$ and $-$ represent two species of atoms, which are trapped at the nodes of left circular polarization wells and at those of the right circular polarization wells, respectively, shown in Figure 10. As discussed previously, these two species of atoms are stored alternatively along the z-axis.

To entangle two neighboring atoms, they are first moved close by tuning the relative angle θ. A monochromatic light beam is then applied to the two atoms which are now at the same site. The light beam, also called a *catalysis field*, is carefully turned on so that it will excite the dipoles in the two atoms and induce dipole-dipole interaction between them. This interaction is strong and only a short time is needed to complete a 2-bit entanglement gate. Although neighboring pairs among the other atoms may also be moved close during this time period, the interaction between them is relatively weak and its effect is negligible. For the two atoms involved, their Hamiltonian includes three parts, the unperturbed Hamiltonian, the atom-light interaction Hamiltonian, and the dipole-dipole interaction Hamiltonian.

The last two can be obtained as

$$
H_{AL} = -\hbar \left(\Delta + i\frac{\Gamma}{2} \right) (D_1^\dagger \cdot D_1 + D_2^\dagger \cdot D_2)
$$

$$
- \frac{\hbar\Omega}{2} (D_1^\dagger \cdot \boldsymbol{e}_L(\boldsymbol{r}_1) + D_2^\dagger \cdot \boldsymbol{e}_L(\boldsymbol{r}_2) + h.c.),
$$

$$
H_{dd} = V_{dd} - i\frac{\hbar\Gamma_{dd}}{2} = -\frac{\hbar\Gamma}{2} (D_2^\dagger \cdot \overleftrightarrow{\boldsymbol{T}}(k_L r)
$$

$$
\cdot D_1 + D_1^\dagger \cdot \overleftrightarrow{\boldsymbol{T}}(k_L r) \cdot D_2), \tag{3.58}
$$

where the tensor $\overleftrightarrow{\boldsymbol{T}}$ describes the strength of the interaction between the two atoms at distance r, $D_{1,2}^\dagger$ is the dipole raising operator, $\boldsymbol{e}_L$ is the polarization vector of the laser field, and V_{dd} is the dipole-dipole energy level shift. Here the anti-Hermitian part, $i\Gamma_{dd}$, i.e., $(i\Gamma_{dd})^\dagger = -i\Gamma_{dd}$, determines the spontaneous emission process. In the above, for a simple two-level atom, the dipole raising operator $D^\dagger$ can be defined as $|e\rangle\langle g|$, which appears in Equation (3.22). However, it could be much complicated when the atom has multiple hyperfine sublevels. Let F and F' denote the hyperfine sublevels for the ground and excited states, respectively. A definition of the operator $D^\dagger$ can be given by

$$
D^\dagger = \sum_{F'} \frac{P_{F'} \boldsymbol{d} P_F}{\langle J' | |\boldsymbol{d}| |J\rangle}
$$

where P_F and $P_{F'}$ are projectors onto the ground and excited manifolds, respectively, and $\boldsymbol{d}$ is the electric dipole moment operator.

The catalysis field is tuned on carefully near the $|S_{1/2}, F_\uparrow\rangle \to |P_{3/2}\rangle$ resonance with small detuning compared with the ground state hyperfine splitting, shown in Figure 10. Thus, dipole is excited for an atom only

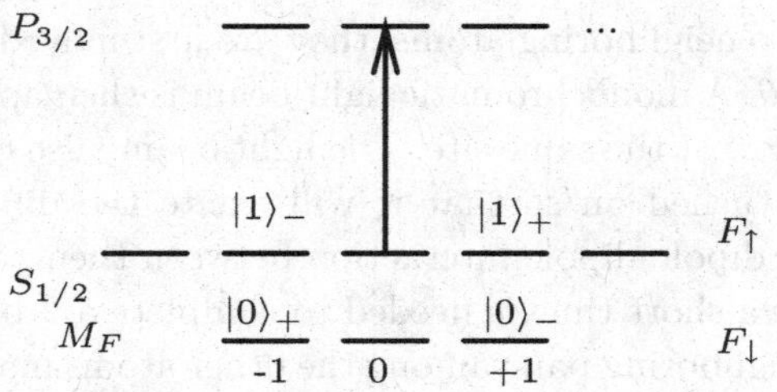

Fig. 10. A schematic of the energy levels used in the scheme with induced dipole-dipole interaction (not in scale). The logic basis $|0\rangle$ and $|1\rangle$ are represented by two ground $S_{1/2}$ hyperfine structures. The atom used has nuclear spin 3/2 ($I = 3/2$). The catalysis laser is tuned near resonant to the $|S_{1/2}, F\uparrow\rangle \to |P_{3/2}\rangle$ transition.

when the atom is in the $|F_\uparrow\rangle$ state, or logic state $|1\rangle$, and the dipole-dipole interaction is strong only when both atoms are in the logic state $|1\rangle$. When written in the matrix form in terms of the ordered basis $\{|00\rangle, |01\rangle, |10\rangle,$ and $|11\rangle\}$, the effective Hamiltonian caused by the induced dipole-dipole interaction is simply

$$V_{dd} = \begin{bmatrix} 0 & 0 & 0 & \\ 0 & 0 & 0 & \\ 0 & 0 & 0 & \\ 0 & 0 & & \nu_{dd} \end{bmatrix}, \qquad (3.59)$$

where ν_{dd} is the energy shift resulted from the interaction. This leads to a *controlled phase gate* on the two atoms:

$$U(\alpha) = e^{-iV_{dd}t/\hbar} = \begin{bmatrix} 1 & & & \\ & 1 & & \\ & & 1 & \\ & & & e^{-i\alpha} \end{bmatrix}, \qquad (3.60)$$

where $\alpha = \frac{\nu_{dd}t}{\hbar}$ and t is the time duration of the interaction.

4. Universality of Quantum Gates

In this section, we intend to provide a unified derivation of the universality of two specific physical devices of quantum computation. The first is a cavity-QED system and the second is a coupled pair of quantum dots. More specifically, we will consider the generation of one specific 2-bit gate for each system, each of which together with the collection of all 1-bit gates is known to be universal for quantum computation. This is a famous theorem due to D. DiVincenzo [16]. Here, instead, we quote the following two theorems due to J.-L. Brylinski and R. Brylinski [7], which are of somewhat more mathematical flavor.

Theorem 4.1 ([7, Theorem 4.1]). *Let V be a given 2-bit gate. Then the following are equivalent*:

(i) *The collection of all 1-bit gates together with V is universal*;
(ii) *V is not a tensor product of two 1-bit gates either with or without a swapping operation.*

Theorem 4.2 ([7, Theorem 4.4]). *The collection of all the 1-bit gates*

$$U_{\theta,\phi} \equiv \begin{bmatrix} \cos\theta & -ie^{-i\phi}\sin\theta \\ -ie^{i\phi}\sin\theta & \cos\theta \end{bmatrix}, \qquad 0 \le \theta, \phi \le 2\pi,$$

together with any 2-bit gate

$$Q_\eta \equiv \begin{bmatrix} 1 & 0 & 0 & 0 \\ 0 & 1 & 0 & 0 \\ 0 & 0 & 1 & 0 \\ 0 & 0 & 0 & e^{i\eta} \end{bmatrix}, \qquad 0 \leq \eta \leq 2\pi,$$

where $\eta \not\equiv 0 (\mathrm{mod} 2\pi)$, is universal.

Here, for the cavity-QED system the 2-bit gate is a quantum phase gate (QPG), while for the quantum dot system the specific 2-bit gate is the square root of swap (SWAP, U_{sw}). We will display an explicit conjugation between the Hamiltonians governing the evolution of the two systems (modulo an overall phase). We have found that the venerable *magic basis matrix* can be taken to be this conjugation.

We need to emphasize that that no claim is being made regarding the two systems being isomorphic under all circumstances. Indeed, the Hamiltonians describing the mechanisms for generating one qubit gates in the two systems are not conjugate (even if one were to ignore overall phases).

Throughout this section, we work with atomic units $\hbar = 1$.

4.1. *The Hamiltonian for the cavity-QED system*

Following [10, pp. 123–130] and Subsection 3.3, we consider a three level atom which has been injected into a cavity. Recall that the total Hamiltonian is given by (3.35). As given in (3.42), in the large detuning limit the effective Hamiltonian becomes

$$\widetilde{H}_2 = -\frac{g^2}{\Delta}(|\alpha\rangle\langle\alpha|aa^\dagger - |\beta\rangle\langle\beta|a^\dagger a), \quad \text{(setting } \hbar = 1 \text{ in (3.42)).}$$

Then it can be shown that the space $\widetilde{V} = \mathrm{span}\{|0,0\rangle, |0,1\rangle, |1,0\rangle, |1,1\rangle\}$ is an invariant subspace for $\widetilde{H}_2$. Furthermore, $\widetilde{H}_2$ has the matrix representation

$$\widetilde{H}_2 = \mathrm{diag}(0, 0, 0, E_{11}); \quad \text{cf. (3.44)},$$

where $E_{11} = \frac{g^2}{\Delta}$. We note that this Hamiltonian has two eigenvalues — one with multiplicity 3 and the second with multiplicity 1. For reasons which will become clear shortly it is convenient to rewrite $\widetilde{H}_2$ as

$$\widetilde{H}_2 = E_{11}\left(-\frac{1}{4}\widehat{H}_2 + \frac{1}{4}I_4\right).$$

Here $\widehat{H}_2 = \text{diag}(1,1,1,-3)$. Note that $\widehat{H}_2$ is traceless. Therefore,

$$e^{-i\widetilde{H}_2 T} = e^{-i\frac{TE_{11}}{4}} e^{i\frac{E_{11}T}{4}\widehat{H}_2}.$$

Thus, up to an overall phase of $e^{-i\frac{TE_{11}}{4}}$, the evolution of $\widetilde{H}_2$ and $\widehat{H}_2$ agree. The former is precisely the quantum phase gate $\text{diag}(1,1,1,e^{i\eta})$, where $\eta = -E_{11}T$.

4.2. *The coupled quantum dot Hamiltonian*

In [9] Burkard *et al.* consider coupled quantum dots as a device for quantum computation. Since our interest is in generating a universal 2-bit quantum gate, only the following Hamiltonian will be considered here. The list of assumptions, including the fact that the only control field operative is the time dependent exchange constant, are detailed in [10, pp. 320–324], and they will not be repeated elsewhere later for brevity:

$$H_1 = \frac{\omega(t)}{2}\widehat{H}_1,$$

where $\omega(t) = 2J_{12}(t)$, with $J_{12}(t)$ is the time-dependent exchange constant, which in turn is $\frac{4t_{12}^2(t)}{u}$. Here $t_{12}(t)$ is the tunable tunneling matrix element between the first and second quantum dots, while u is the charging energy of a single dot. The Hamiltonian H_1 is given by $\widehat{H}_1 = \sigma_x \otimes \sigma_x + \sigma_y \otimes \sigma_y + \sigma_z \otimes \sigma_z$, where σ_x, σ_y and σ_z are the standard Pauli matrices mentioned earlier in Section 3.

Of importance is the fact that $\widehat{H}_1$'s spectrum consists of the eigenvalues 1 (with triple multiplicity) and -3 with multiplicity one. Thus $i\widehat{H}_1$ and $i\widehat{H}_2$ are unitarily equivalent. It would seem that one would have to diagonalize both $\widehat{H}_i, i = 1,2$ to arrive at an explicit conjugation. However, we will see that this is not the case.

For the moment let us first observe that the unitary evolution operator at time T is

$$U(T) = \exp\left(-\frac{i}{2}\int_0^T \omega(s)ds\widehat{H}_1\right).$$

From a variety of viewpoints $\widehat{H}_1$ is easy to exponentiate. This yields

$$U(T) = e^{i\phi}\left[\cos(2\phi)I_4 - i\sin(2\phi)\frac{I + \widehat{H}_1}{2}\right],$$

where $\phi = \frac{1}{2}\int_0^T \omega(s)ds$. Since the swap gate is $U_{sw} = \frac{I+\widehat{H}_1}{2}$, we obtain the universal 2-bit gate U_{sw}, up to an overall phase, by picking 2ϕ to be an odd-integer multiple of $\frac{\pi}{2}$. Hence, choosing ϕ to be an odd-integer multiple of $\frac{\pi}{8}$ would produce, up to an overall phase, the square root of U_{sw}.

4.3. *An explicit conjugation*

Since $i\widehat{H}_i, i = 1, 2$ are unitarily equivalent, there exists $M \in U(4)$ with $M^*(i\widehat{H}_1)M = i\widehat{H}_2$. Hence by choosing $\frac{E_{11}T}{4} = -\frac{1}{2}\int_0^T \omega(s)ds$, we also obtain that the swap and phase gates are (up to a phase factor) unitarily equivalent to one another, via the same M. Thus, it is very pertinent to obtain an explicit expression for one such M. As mentioned above, one can do this without resorting to any spectral calculations. The key to this is the existence of two Type I Cartan decompositions of $su(4)$.

In the interest of keeping the discussion below self-contained, some of the essentials of the theory of Cartan decompositions will be briefly recalled. See [24] for a comprehensive account and also [8] and the references therein which highlight the importance of this notion for quantum information processing.

The theory of Cartan decompositions makes sense for any semisimple Lie algebra, [24], but we will restrict ourselves to the case that the Lie algebra in question is $\mathfrak{g} = su(N)$, the Lie algebra of $N \times N$ anti-hermitian matrices of zero trace.

A preliminary Cartan decomposition of $\mathfrak{g} = su(N)$ is a vector space direct sum decomposition $\mathfrak{g} = \mathfrak{h} \oplus \mathfrak{p}$ (with respect to the Killing form - to be defined shortly), satisfying

- $[\mathfrak{h}, \mathfrak{h}] \subseteq \mathfrak{h}$, i.e., $\mathfrak{h}$ is a Lie subalgebra.
- $[\mathfrak{p}, \mathfrak{p}] \subseteq \mathfrak{h}$.
- $[\mathfrak{h}, \mathfrak{p}] \subseteq \mathfrak{p}$.

In this direct sum decomposition, $\mathfrak{h}$ and $\mathfrak{p}$ are the orthogonal complements of each other with respect to the Killing form. The Killing form on $\mathfrak{g}$ is the non-degenerate bilinear form $K(X, Y) = \text{Tr}[\text{ad}(X)\text{ad}(Y)], X, Y \in \mathfrak{g}$. Here $\text{ad}(X)$ is the linear map from $\mathfrak{g}$ to itself defined by

$$\text{ad}(X)(Z) = [X, Z].$$

Corresponding to this preliminary Cartan decomposition is a decomposition of the Lie group $SU(N)$. More precisely, every $G \in SU(N)$ can be factorized as HP, where H is the exponential of an element in $\mathfrak{h}$ and P an exponential of an element in $\mathfrak{p}$.

Now, the conditions $[\mathfrak{p}, \mathfrak{p}] \subseteq \mathfrak{h}$ and $\mathfrak{h} \cap \mathfrak{p} = \{0\}$ ensures that any Lie subalgebra contained in $\mathfrak{p}$ is necessarily abelian. With this observation we are ready to define a Cartan decomposition of $\mathfrak{g}$. It consists of

- a preliminary Cartan decomposition of $\mathfrak{g}$;
- a maximal abelian Lie sublagebra $\mathfrak{a}$ contained in $\mathfrak{p}$, called a Cartan subalgebra.

Corresponding to a Cartan decomposition of $\mathfrak{g}$, there is a decomposition of $SU(N)$, i.e., every $G \in SU(N)$ can be factorized as $H_1 A H_2$, where $H_i, i = 1, 2$ are exponentials of elements in $\mathfrak{h}$ and A is an exponential of an element in $\mathfrak{a}$.

Cartan decompositions are not unique. There are two levels of non-uniqueness, namely, the following:

- Even for a fixed preliminary decomposition of $\mathfrak{g}$ (i.e., $\mathfrak{h}, \mathfrak{p}$ are fixed) there may be many Cartan subalgebras $\mathfrak{a}_i$. However, all such Cartan subalgebras are conjugate via elements of $e^{\mathfrak{h}}$, i.e., given two Cartan subalgebras $\mathfrak{a}_i, i = 1, 2$ contained in $\mathfrak{p}$, there is an element $H \in SU(N)$ which is an exponential of an element of $\mathfrak{h}$ such that $\mathfrak{a}_2 = H^* \mathfrak{a}_1 H$. This implies that the dimension of all Cartan subalgebras corresponding to a given preliminary decomposition are the same. This dimension is called the *rank* of the Cartan decomposition.
- There can be several preliminary Cartan decompositions. We will identify two Cartan decompositions $(\mathfrak{h}_1, \mathfrak{p}_1, \mathfrak{a}_1)$ and $(\mathfrak{h}_2, \mathfrak{p}_2, \mathfrak{a}_2)$ of $su(N)$ if there exists a $G \in SU(N)$ satisfying $G^* \mathfrak{h}_1 G = \mathfrak{h}_2, G^* \mathfrak{p}_1 G = \mathfrak{p}_2$ and $G^* \mathfrak{a}_1 G = \mathfrak{a}_2$. It is known that up to such identifications there are precisely three types of Cartan decompositions of $su(N)$. These are called Type I, Type II and Type III. It also known that two Cartan decompositions are of the same type if and only if their ranks are the same.

We will not dwell on what precisely these three types are, except to mention that the rank of the Type I decompositions is $N - 1$.

We now discuss the two Type I Cartan decompositions of $su(4)$ pertinent to the material in the previous section. These are:

- $\mathfrak{h}_1 = so(4, R)$ is a Lie subalgebra of $su(4)$. Its orthogonal complement (with respect to the inner product induced by the Killing form) is $\mathfrak{p}_1$, the vector space of purely imaginary matrices in $su(4)$. The space $\mathfrak{a}_1$ of diagonal matrices in $\mathfrak{p}_1$ is a Cartan subalgebra. This yields the so-called standard Type I Cartan decomposition of $su(4)$. The rank of this decomposition is 3.

- Let $\mathfrak{h}_2 = \text{span}\{iI_2 \otimes \sigma_i, i\sigma_j \otimes I_2, i = x, y, z; j = x, y, z\}$. This is a Lie subalgebra of $su(4)$. Its orthogonal complement (with respect to the inner product induced by the Killing form) is $\mathfrak{p}_2$, which is the real span of matrices of the form $\{i\sigma_j \otimes \sigma_k, I_4, j, k = x, y, z\}$. The space $\mathfrak{a}_2$ of matrices which lie in the real span of $\{i\sigma_j \otimes \sigma_j, j = x, y, z\}$ is a Cartan subalgebra in $\mathfrak{h}_2$. Since the rank of this Cartan decomposition is also 3, this is also a Type I Cartan decomposition of $su(4)$.

Hence there must exist an $M \in SU(4)$ which renders these two Type I Cartan decompositions of $su(4)$ conjugate to each other. Furthermore, this M is a matrix which will make the $\mathfrak{a}_i$ conjugate to each other and hence simultaneously diagonalize all the commuting matrices in $\mathfrak{a}_2$. In particular, since $i\widehat{H}_1 \in \mathfrak{a}_2$, it must be unitarily equivalent by that M to a matrix in $\mathfrak{a}_1$. Since all the matrices in $\mathfrak{a}_1$ are diagonal, this matrix has to be the diagonal form of $\widehat{H}_1$. It is easy to see that this must necessarily be $i\widehat{H}_2$. Due to its ubiquity in quantum information theory, it has been known since the very early days of this field that one such M is precisely the magic basis matrix,

$$M = \frac{1}{\sqrt{2}} \begin{pmatrix} 1 & 0 & 0 & i \\ 0 & i & 1 & 0 \\ 0 & i & -1 & 0 \\ 1 & 0 & 0 & -i \end{pmatrix}.$$

Thus $M^*(i\widehat{H}_1)M = i\widehat{H}_2$.

Of course, this is not the sole M which renders $i\widehat{H}_i, i = 1, 2$ equivalent. Indeed, it is not even the sole M which renders the above Type I Cartan decompositions of $su(4)$ equivalent. An $M \in SU(4)$ renders the above Cartan decompositions equivalent if and only if $MM^T = \psi(\sigma_y \otimes \sigma_y)$, with ψ a fourth root of unity [8].

Note the above choice of setting $\frac{E_{11}T}{4} = -\frac{1}{2}\int_0^T \omega(s)ds$, may not always be possible if the control fields in both the cavity-QED system and the quantum-dot system are always required to be of the same sign. However, in this case by choosing $\frac{1}{2}\int_0^T \omega(s)ds$ and $\frac{E_{11}T}{4}$ the same, we still obtain gates which are, up to a phase factor, the swap and the phase gate respectively.

Acknowledgments

G. Chen and V. Ramakrishna are happy to acknowledge the support from the Institute for Mathematical Sciences, National University of Singapore where they both visited during the summer of 2008. Interactions and discussions with other visitors of the MHQP program were invaluable.

References

1. P. Ball, http://www.nature.com/nature/journal/v440/n7083/pdf/440398a.pdf
2. P. Benioff, "The computer as a physical system: a microscopic quantum mechanical Hamiltonian model of computers as represented by Turing machines," *J. Stat. Phys.*, 22, pp. 563–591, 1980.
3. C. H. Bennett, *Int. J. Theor. Phys.*, 21, 905, 1982.
4. I. Bloch, "Quantum coherence and entanglement with ultracold atoms in optical lattices," *Nature*, 453, pp. 1016–1022, 2008.
5. G. K. Brennen, C. M. Caves, P. S. Jessen and I. H. Deutsch, "Quantum logic gates in optical lattices," *Phys. Rev. Lett.*, 82, pp. 1060–1063, 1999.
6. G. K. Brennen, I. H. Deutsch and P. S. Jessen, "Entangling dipole-dipole interaction for quantum logic with neutral atoms," *Phys. Rev. A*, 61, 062309, 2000.
7. J.-L. Brylinski and R. Brylinski, "Universal quantum gates," in *Mathematics of Quantum Computation*, R. Brylinski and G. Chen (eds.), Chapman & Hall/CRC, Boca Rato, Florida, pp. 101–116, 2002.
8. S. Bullock, G. Brennen and D. O'Leary, "Time reversals and n-qubit canonical decompositions," *J. Math. Phys.*, 46, 062104, 2005.
9. G. Burkard, D. Loss and D. Divincenzo, "Coupled quantum dots as quantum gates," *Phys. Rev. B*, 59, pp. 2070–2078, 1999.
10. G. Chen, D. A. Church, B.-G. Englert, C. Henkel, B. Rohwedder, M. O. Scully and M. S. Zubairy, *Quantum Computing Devices, Principles, Designs and Analysis*, Chapman & Hall/CRC, Boca Raton, Florida, 2006.
11. G. Chen, "Three fundamental principles of quantum computing," in *CMASM-FIM XIV (Computational, Mathematical and Statistical Methods, The Fourteenth International Conference of the Forum for Interdisciplinary Mathematics)* program booklet, Chennai, India, Jan. 6-8, 2007, pp. 41–52.
12. J. F. Clauser, M. A. Horne, A. Shimony and R. A. Holt, "Proposed experiment to test local hidden-variable theories," *Phys. Rev. Lett.*, 23, pp. 880–884, 1969.
13. C. Cohen-Tannoudji, "Atomic motion in laser light," in Les Houches, Session LIII, *Fundamental Systems in Quantum Optics*, J. Dailibard, J. Raimond and J. Zinn-Justin (eds.), Elsevier Science Publisher B.V., pp. 1–164, 1990.
14. D. Deutsch, "Quantum theory, the Church–Turing principle, and the universal quantum computer," *Proc. Royal Soc. London*, A400, pp. 97–117, 1985.
15. D. Deutsch, "Quantum computational networks," *Proc. Royal Soc. London*, A425, p. 73, 1989.
16. D. P. DiVincenzo, "Two-bit quantum gates are universal for quantum computation," *Phys. Rev. A*, 51, pp. 1015–1022, 1996.
17. A. Einstein, B. Podolsky and N. Rosen, *Phys. Rev.*, 47, p. 777, 1935.
18. R. P. Feynman, "Simulating physics with computers," *Int. J. Theor. Phys.*, 21, pp. 467–488, 1982.
19. B. Friedrich and D. Herschbach, "Stern and Gerlach: How a bad cigar helped reorient atomic physics," *Phys. Today*, 56, pp. 53–59, 2003.

20. S. Goudsmit, *Z. Physik*, 32, p. 111, 1925.
21. S. Goudsmit and R. de L. Kronig, *Naturwissenschaften*, 13, p. 90, 1925; *Verhandelingen Koninklijke Akademie van Wetenschappen*, 34, p. 278, 1925.
22. D. Jaksch, H. J. Briegel, J. I. Cirac, C. W. Gardiner and P. Zoller, "Entanglement of atoms via cold controlled collisions," *Phys. Rev. Lett.*, 82, pp. 1975–1978, 1999.
23. Y.-H. Kim, R. Yu, S. P. Kulik, Y. Shih and M. O. Scully, *Phys. Rev. Lett.*, 84, p. 1, 2000.
24. A. Knapp, *Lie Groups: Beyond an Introduction*, Birkhäuser, Basel, 2001.
25. R. Landauer, *IBM J. Res. Develp.*, 5, p. 183, 1961.
26. Z. Meglicki, *Quantum Computing without Magic: Devices*, MIT Press, Cambridge, MA, 2008.
27. Y. Miroshnychenko, W. Alt, I. Dotsenko, L. Förster, M. Khudaverdyan, D. Meshede, D. Schrader and A. Rauschenbeutel, "An atom-sorting machine," *Nature*, 442, p. 151, 2006.
28. M. A. Neilsen and I. L. Chuang, *Quantum Computation and Quantum Information*, Cambridge University Press, Cambridge, U.K., 2000.
29. M. B. Plenio and V. Vitelli, "The physics of forgetting: Landauer's erasure principle and information theory," *Contemporary Phys.*, 42, pp. 25–60, 2001.
30. D. Schrader, I. Dotsenko, M. Khudaverdyan, Y. Miroshnychenko, A. Rauschenbeutel and D. Meschede, "Neutral atom quantum register," *Phys. Rev. Lett.*, 93, 150501, 2004.
31. E. Schrödinger, *Naturwissenschaften*, 23, pp. 807, 823 and 844, 1935; English translation by J. D. Trimmer, *Proc. Amer. Philos. Soc.*, 124, p. 323, 1980.
32. M. O. Scully and K. Drühl, *Phys. Rev. A*, 25, p. 2208, 1982.
33. M. O. Scully, B.-G. Englert and H. Walther, *Nature (London)*, 351, p. 111, 1991.
34. M. Scully and M. S. Zubairy, *Quantum Optics*, Cambridge University Press, Cambridge, U.K., 1997.
35. R. Scully and M. O. Scully, *The Demon and the Quantum*, book to appear, 2006.
36. P. Shor, "Polynomial-time algorithms for prime factorization and discrete logarithms on a quantum computer," *SIAM J. Comput.*, 26, pp. 1481–1509, 1997.
37. L. Szilard, *Z. Phys.*, 53, p. 840, 1929.
38. G. E. Uhlenbeck and S. Goudsmit, *Naturwissenschaften*, 47, p. 953, 1925.
39. http://en.wikipedia.org/wiki/CHSH_inequality.
40. Z. Zhang and G. Chen, "Mathematical formulation of atom trap quantum gates," *Contemporary Math.*, 482, pp. 1–22, 2009.
41. X. Y. Zou, L. J. Wang and L. Mandel, *Phys. Rev. Lett.*, 67, p. 318, 1991; see also P. G. Kwiat, A. M. Steinberg and R. Y. Chiao, *Phys. Rev. A*, 47, p. 7729, 1992.

DYNAMICS OF MIXED CLASSICAL-QUANTUM SYSTEMS, GEOMETRIC QUANTIZATION AND COHERENT STATES

Hans-Rudolf Jauslin* and Dominique Sugny[†]

Institut Carnot de Bourgogne UMR 5209 CNRS
Université de Bourgogne
BP 47870, 21078 Dijon, France
*E-mails: *jauslin@u-bourgogne.fr*
[†] *dominique.sugny@u-bourgogne.fr*

We describe quantum and classical Hamiltonian dynamics in a common Hilbert space framework, that allows the treatment of mixed quantum-classical systems. The analysis of some examples illustrates the possibility of entanglement between classical and quantum systems. We give a summary of the main tools of Berezin-Toeplitz and geometric quantization, that provide a relation between the classical and the quantum models, based essentially on the selection of a subspace of the classical Hilbert space. Coherent states provide a systematic tool for the inverse process, called dequantization, that associates a classical Hamiltonian system to a given quantum dynamics through the choice of a complete set of coherent states.

1. Introduction

In order to describe the dynamics of a bipartite system, in which one part is quantum and the other one is classical, we have to describe the classical and quantum systems in a common mathematical framework. The construction is guided by the idea that we can consider the classical system as a limit, and as a particular case, of a quantum system. The framework we consider is provided by the Koopman-von Neumann [1, 2] representation of classical mechanics combined with a quantization procedure of the Berezin-Toeplitz type.

This construction is to be distinguished from models in which the coupling between a classical system (e.g. a classical electromagnetic field) and a quantum system composed of many identical atoms is described by a coupling involving the average of some quantum observable (e.g. the electric dipole moment). This type of effective models is useful e.g. to describe the induced classical field produced by a large number of atoms excited by the initial electromagnetic wave. It is implicitly assumed that the quantum average provides an effective description of the collective retroaction of the atoms on the classical system. The models we discuss in this article are meant to describe the dynamics of a single quantum system in interaction with a single classical object. A prototypical example is a model for the Stern-Gerlach experiment in which the motion of the center of mass of the atom is treated classically and the spin is treated quantum mechanically.

In Section 2 we summarize the mathematical framework of quantum dynamics. In Section 3 we describe a formulation of classical mechanics in a Hilbert space framework, on the basis of the formulation of Koopman and von Neumann. Once classical and quantum dynamics are formulated in the same mathematical framework, in Section 4 we describe the formalism to treat mixed systems, in which one part is classical and another one is quantum. We illustrate this formalism with a simplified model for the Stern-Gerlach experiments, in which the motion of the center of mass is considered as classical and the spin quantum mechanical. Section 5 presents with a minimum of mathematical formalism some of the main ideas of geometric and Berezin-Toeplitz quantization. This allows to establish a well-defined relation between classical and quantum models. The main idea is that one can obtain quantum models from the Koopman-von Neumann representation of classical dynamics in a Hilbert space just by selecting a suitable subspace of the classical Hilbert space. In Section 6 we describe the inverse process, called dequantization, in which starting with a given quantum system and a set of coherent states one constructs a classical phase space and Hilbert space, a corresponding polarization subspace, classical observables and a classical dynamics.

2. Hilbert Space Framework of Quantum Dynamics

A quantum dynamical system is defined by the following elements:

   (i) A separable Hilbert space $\mathcal{H}$.

  (ii) An algebra of observables $\mathcal{A}$ and a representation $\rho(\mathcal{A})$ as linear operators on the Hilbert space $\mathcal{H}$. The physical observables are given by the self-adjoint elements.

 (iii) The dynamics, in the Heisenberg representation, is defined by a derivation operator $D_{\hat{H}}$, that acts on the algebra $\rho(\mathcal{A})$. The derivation is constructed from the Hamiltonian $\hat{H}$, which is a particular self-adjoint element of the algebra $\mathcal{A}$. The derivation $D_{\hat{H}}$ can be expressed in terms of the commutator with a linear operator $\hat{H}$ acting on the Hilbert space $\mathcal{H}$: $D_{\hat{H}} = i[\frac{1}{\hbar}\hat{H}, \cdot]$. The Heisenberg equation for an observable $\hat{A}$ can be written as

$$\frac{\partial \hat{A}}{\partial t} = D_{\hat{H}}(\hat{A}) = i\left[\frac{1}{\hbar}\hat{H}, \hat{A}\right].$$

(iv) The Hamiltonian $\hat{H}$ defines also the dynamics of the states $\psi \in \mathcal{H}$ in the Schrödinger picture, by

$$i\frac{\partial \psi}{\partial t} = \frac{1}{\hbar}\hat{H}\psi.$$

We remark that, in order to prepare a common notation with the classical systems, all the dependence on Planck's constant $\hbar$ is attached to the operators.

3. Hilbert Space Framework of Classical Dynamics

The Koopman-von Neumann [1–3] formalism is a representation of classical dynamics on a phase space $\mathcal{M}$ in terms of a Hilbert space of square integrable functions $\mathcal{L}_K := L_2(\mathcal{M}, d\mu)$, with respect to a measure $d\mu$. The classical observables, which are differentiable functions $f : \mathcal{M} \to \mathbb{C}$ are represented on $\mathcal{L}_K$ by a commutative algebra of multiplication operators M_f: For $\xi \in \mathcal{L}_K$

$$(M_f \xi)(\underline{z}) := f(\underline{z})\xi(\underline{z}).$$

In order to simplify the notation we will sometimes write f instead of M_f.

3.1. *Koopman-Schrödinger representation*

The classical time evolution is determined by a Hamilton function $H_{cl}(\underline{z})$, where we denote the local coordinates by $\underline{z} = (p, q) \in \mathcal{M}$, $p = p_1, \ldots, p_d$, $q = q_1, \ldots, q_d$. The classical Hamiltonian flow, denoted by $\underline{z}(t) = \Phi_t(\underline{z}_0)$, which satisfies the classical Hamilton equations

$$\frac{dp}{dt} = -\frac{\partial H_{cl}}{\partial q}, \qquad \frac{dq}{dt} = \frac{\partial H_{cl}}{\partial p}, \tag{3.1}$$

which can be written as

$$\frac{d\Phi_t(\underline{z}_0)}{dt} = J \, \nabla H_{cl}(\Phi_t(\underline{z}_0)); \qquad J = \begin{pmatrix} 0 & -1 \\ 1 & 0 \end{pmatrix}, \qquad \nabla = \begin{pmatrix} \frac{\partial}{\partial p} \\ \frac{\partial}{\partial q} \end{pmatrix}$$

can be used to define a unitary flow on the Koopman Hilbert space [1, 2, 39]: $U^K(t, t_0)\xi_0(\underline{z}) := \xi_0(\Phi_{-(t-t_0)}(\underline{z}))$. We will use a slightly different unitary flow, which we will call Koopman-Schrödinger flow, that includes a time-dependent phase:

$$\xi(\underline{z}, t) := U^{KS}(t, t_0)\xi_0(\underline{z}) := e^{-i\delta(\underline{z}, t-t_0)}\xi_0(\Phi_{-(t-t_0)}(\underline{z})), \quad \xi(\underline{z}, 0) = \xi_0(\underline{z}), \tag{3.2}$$

with

$$\delta(\underline{z}, t - t_0) = \frac{1}{\hbar}\left((t - t_0)\, H_{cl}(\underline{z}) + \int_{t_0}^{t} dt'\, \Lambda_{H_{cl}}(\Phi_{t'-(t-t_0)}(\underline{z}))\right), \tag{3.3}$$

where the function $\Lambda_{H_{cl}}(p, q)$ will be chosen as [4–7]

$$\Lambda_{H_{cl}}(p, q) = -\frac{1}{2}\sum_{j}\left(p_j\frac{\partial H_{cl}}{\partial p_j} + q_j\frac{\partial H_{cl}}{\partial q_j}\right). \tag{3.4}$$

The flow (3.2) with the phase (3.3) is called the *prequantum flow* in the framework of *geometric quantization* [4–6]. The first term in (3.3) is called a *dynamical phase* and the second one a *geometrical phase*. The constant $\hbar$ introduced in (3.3), which has the dimension of an action ([time]·[energy]) is necessary in order to have dimensionless numbers in the exponential. Its numerical value has to be determined by comparison between the predictions of the quantum or of the mixed models with the corresponding experimental measurements.

The interpretation of the classical wave function $\xi(p, q, t)$ is that $\rho d\mu := |\xi(p, q, t)|^2 d\mu$ gives the probability density at time t for a particle to have a momentum p and a position q. This interpretation comes from the fact that ρ is a real positive function that satisfies the Liouville equation $\frac{\partial\rho}{\partial t} = -\{H_{cl}, \rho\}$, which is the equation that describes the time evolution of the probability densities ρ in phase space defined by the flow of Eqs. (3.1). Indeed Eq. (3.2) implies that $\rho(\underline{z}, t) = \rho_0(\Phi_{-(t-t_0)}(\underline{z}))$, with $\rho_0 := |\xi_0|^2$, and thus $\frac{\partial\rho}{\partial t} = -\{H_{cl}, \rho\}$, where the brackets $\{\,\cdot\,\}$ denote the Poisson brackets defined by $\{h, f\} := \sum_j \frac{\partial h}{\partial p_j}\frac{\partial f}{\partial q_j} - \frac{\partial h}{\partial q_j}\frac{\partial f}{\partial p_j}$.

For a purely classical system this global phase $\delta(\underline{z}, t)$ is thus irrelevant since it is the square of the absolute value $|\xi(\underline{z}, t)|^2$ that gives all the physically relevant quantities. We will see that the analogue of the phase term plays an essential role in the description of quantum systems and of mixed classical-quantum systems.

By Stone's theorem [39], this unitary flow has a selfadjoint generator $G_{H_{cl}}$, i.e. it can be written as

$$U^{KS}(t,t_0) = U^{KS}(t-t_0) = e^{-i(t-t_0)G_{H_{cl}}}. \qquad (3.5)$$

The dynamics of the vectors of the Hilbert space satisfies therefore the following Schrödinger type equation, which we will call the Koopman-Schrödinger equation:

$$i\frac{\partial\xi}{\partial t} = G_{H_{cl}}\xi. \qquad (3.6)$$

The generator $G_{H_{cl}}$ can be written explicitly as the sum of three operators

$$G_{H_{cl}} = \tilde{M}_{H_{cl}} + \tilde{M}_{\Lambda_{H_{cl}}} + \tilde{X}_{H_{cl}}, \qquad (3.7)$$

or explicitly,

$$G_{H_{cl}} = \frac{1}{\hbar}H_{cl} - \frac{1}{2\hbar}\sum_j\left(p_j\frac{\partial H_{cl}}{\partial p_j} + q_j\frac{\partial H_{cl}}{\partial q_j}\right) - i\sum_j\frac{\partial H_{cl}}{\partial p_j}\frac{\partial}{\partial q_j} - \frac{\partial H_{cl}}{\partial q_j}\frac{\partial}{\partial p_j}, \qquad (3.8)$$

where $\tilde{M}_{H_{cl}} := \frac{1}{\hbar}M_{H_{cl}}$ is the multiplication operator with the Hamilton function, $\tilde{M}_{\Lambda_{H_{cl}}} := \frac{1}{\hbar}M_{\Lambda_{H_{cl}}}$ is the multiplication operator with $\frac{1}{\hbar}\Lambda_{H_{cl}}(p,q)$, and $\tilde{X}_{H_{cl}} := -iX_{H_{cl}}$ is the vector field associated to the Hamilton function H_{cl}, i.e. the differential operator

$$\tilde{X}_{H_{cl}} := -iX_{H_{cl}} = -i\sum_j\frac{\partial H_{cl}}{\partial p_j}\frac{\partial}{\partial q_j} - \frac{\partial H_{cl}}{\partial q_j}\frac{\partial}{\partial p_j} \equiv -i\{H_{cl},\,\cdot\,\}.$$

The dynamics of the classical system defined by Eq. (3.6) is called *pre-quantum dynamics*, since, as we we will discuss in Section 5.3, it is the starting point for the construction of the quantum dynamics in the framework of geometric quantization [4–6]. The motivation for the addition of the terms $\tilde{M}_{H_{cl}}$ and $\tilde{M}_{\Lambda_{H_{cl}}}$ in (3.8) is:

(i) The term $\tilde{M}_{H_{cl}}$ is added in order to have $G_{f(\underline{z})=1} = \frac{1}{\hbar}\mathbb{1}$. If we had only $G_{H_{cl}} = \tilde{X}_{H_{cl}}$ it would lead to $G_{f(\underline{z})=1} = 0$, which would not allow a systematic construction of the dynamics of mixed systems.

(ii) $\tilde{X}_{H_{cl}}$ satisfies $\tilde{X}_{\{f,g\}} = i[\tilde{X}_f, \tilde{X}_g]$, but this property is not satisfied by $\tilde{M}_{H_{cl}} + \tilde{X}_{H_{cl}}$. The motivation of the choice (3.4) for $\Lambda_{H_{cl}}$ is that the operator $\nabla_{H_{cl}} := M_{\Lambda_{H_{cl}}} + X_{H_{cl}}$ can be interpreted as a covariant derivation associated to H_{cl} [4, 7], and as a consequence G_g satisfies

$$G_{\{f,g\}} = i[G_f, G_g].$$

Remark. (1) The action of the Poisson brackets is defined both on the algebra of observables (functions $f \in \mathcal{A}$, and their representation as multiplication operators $\rho(f) = M_f$) and on the (differentiable) vectors of the Koopman Hilbert space $\mathcal{L}_K$.

(2) The operators $M_{H_{cl}}$ and $X_{H_{cl}}$ commute, since

$$[M_{H_{cl}}, X_{H_{cl}}]\xi = H_{cl}\{H_{cl}, \xi\} - \{H_{cl}, H_{cl}\xi\}$$

and $\{H_{cl}, H_{cl}\xi\} = H_{cl}\{H_{cl}, \xi\} + \{H_{cl}, H_{cl}\}\xi = H_{cl}\{H_{cl}, \xi\}$.

However, except for special choices of the Hamiltonian H_{cl}, the operators $M_{\Lambda_{H_{cl}}}$ and $X_{H_{cl}}$ do not commute.

The unitary Koopman-Schrödinger operator (3.2), (3.3) can be written as

$$U^{KS}(t) = e^{-i\delta}e^{-i(t-t_0)X_{H_{cl}}} = e^{-i(t-t_0)G_{H_{cl}}}.$$

Some examples of operators $G_{H_{cl}}$ are given in Table 5.1.

3.2. *Koopman-Heisenberg representation*

One can also define a dynamics on the algebra of observables, which we will call the Koopman-Heisenberg representation, by

$$M_f(t) := (U^K(t))^{-1} M_f U^K(t) = e^{itG_{H_{cl}}} M_f e^{-itG_{H_{cl}}}.$$

Since $U^K(t) = e^{-i\delta}e^{-it\tilde{X}_{H_{cl}}}$, it can be also written as

$$M_f(t) = e^{it\tilde{X}_{H_{cl}}} M_f e^{-it\tilde{X}_{H_{cl}}}.$$

It satisfies the following Koopman-Heisenberg equation

$$\frac{\partial M_f}{\partial t} = i[G_{H_{cl}}, M_f], \tag{3.9}$$

which can be written equivalently as

$$\frac{\partial M_f}{\partial t} = i[\tilde{X}_{H_{cl}}, M_f] = M_{\{H_{cl}, f\}}. \tag{3.10}$$

The last equality can be understood by letting it act on an arbitrary vector $\xi \in \mathcal{L}_K$:

$$i[\tilde{X}_{H_{cl}}, M_f]\xi = i\tilde{X}_{H_{cl}}(M_f\xi) - \tilde{X}_f(iH_{H_{cl}}(\xi))$$
$$= \{H_{cl}, (M_f\xi)\} - M_f(\{H_{cl}, \xi\}) = \{H_{cl}, (f\xi)\} - f\{H_{cl}, \xi\}$$
$$= \{H_{cl}, f\}\xi + f\{H_{cl}, \xi\} - f\{H_{cl}, \xi\}$$
$$= \{H_{cl}, f\}\xi = M_{\{H_{cl}, f\}}\xi.$$

We remark that the Koopman-Heisenberg dynamics does not involve the phase $\delta(\underline{z}, t)$ we introduced in (3.2). The Koopman-Heisenberg equation (3.10) is equivalent to the equation for the time evolution of a classical observable $f(p, q)$

$$\frac{df}{dt} = \{H_{cl}, f\},$$

but written in terms of the corresponding multiplication operator. Thus, the Koopman-Heisenberg equations for the particular observables p and q are equivalent to Hamilton's equations.

We remark that the generator of the dynamics is an operator that does not belong to the algebra of observables, but it is an external derivation acting on them. Indeed for a Hamilton function H_{cl} the dynamics is defined by the derivation $D_{H_{cl}}$, acting on the algebra $\rho(\mathcal{A})$ by

$$D_{H_{cl}}(M_f) := i[X_{H_{cl}}, M_f] \equiv M_{\{H_{cl}, f\}}.$$

This provides a complete framework in which classical mechanics can be considered, from the mathematical point of view, as a particular type of quantum system.

This formulation opens up the possibility to construct models of a quantum system in interaction with a classical one.

4. Dynamics of Mixed Classical-Quantum Systems

We consider a bipartite system composed of a quantum system (Q), defined on a Hilbert space $\mathcal{H}_Q$ and a classical system (K) with a Koopman Hilbert space $\mathcal{L}_K$. The total Hilbert space will be $\mathcal{K} = \mathcal{L}_K \otimes \mathcal{H}_Q$. The observables of the total system will be linear combinations of operators of the form $M_f \otimes \hat{B}$, where $\hat{B}$ is a selfadjoint operator of the quantum system. We first consider the two subsystems without coupling. The evolution of the quantum subsystem is defined by a Hamiltonian $\mathbb{1}_K \otimes \hat{H}_0^Q / \hbar$ and the one of the classical subsystem by $G_{H_{0,cl}} \otimes \mathbb{1}_Q / \hbar$, with $G_{H_{0,cl}} = \tilde{M}_{H_{0,cl}} + \tilde{\Lambda}_{H_{0,cl}} + \tilde{X}_{H_{0,cl}}$, and $H_{0,cl}$ is the uncoupled classical Hamilton function.

The interaction between the classical and the quantum subsytems is determined by observables that are linear combinations of terms of the form

$$M_{g_i} \otimes \hat{\mu}_i / \hbar$$

where $\hat{\mu}_i$ is a self-adjoint operator defined on $\mathcal{H}_Q$ and $g_i = g_i(p, q)$ is a function on the classical phase space.

The corresponding dynamics of the interaction is defined by operators of the form

$$K_{int} = \sum_i G_{g_i} \otimes \hat{\mu}_i / \hbar$$

with $G_{g_i} = \tilde{M}_{g_i} + \tilde{\Lambda}_{g_i} + \tilde{X}_{g_i}$, i.e. $G_{g_i}\xi = \frac{1}{\hbar}g_i\xi - \frac{1}{2\hbar}\left(p\frac{\partial g_i}{\partial p} + q\frac{\partial g_i}{\partial q}\right)\xi - i\{g_i, \xi\}$.

The Schrödinger-Koopman equation for a mixed classical-quantum system can thus be written as

$$i\frac{\partial \psi}{\partial t} = K\psi,$$

with

$$K = \mathbb{1}_{cl} \otimes \hat{H}_{0,Q}/\hbar + G_{H_{0,cl}} \otimes \mathbb{1}_Q/\hbar + \sum_i G_{g_i} \otimes \hat{\mu}_i/\hbar.$$

The corresponding Heisenberg-Koopman equation for the dynamics of an observable of the form

$$B(t) := e^{itK}\left(M_f \otimes \hat{A}\right)e^{-itK}$$

is

$$\frac{\partial B}{\partial t} = i[K, B].$$

4.1. Example: Stern-Gerlach experiment

We consider a simple model of the Stern-Gerlach experiment, in which the motion of the center of mass of the atom is described classically, and the spin as a quantum variable. The total Hilbert space is $\mathcal{K} = \mathcal{L}_2(\mathbb{R}^6, d^3p\, d^3q) \otimes \mathbb{C}^2$. The states $\psi \in \mathcal{K}$ can be represented by

$$\psi = v_+(\vec{p}, \vec{q}) \otimes |+\rangle + v_-(\vec{p}, \vec{q}) \otimes |-\rangle \equiv \begin{pmatrix} v_+(\vec{p}, \vec{q}) \\ v_-(\vec{p}, \vec{q}) \end{pmatrix}.$$

The observable corresponding to the total energy is

$$H = \frac{1}{2m}\vec{p}^{\,2} \otimes \mathbb{1} - \gamma\vec{B}(\vec{q}) \otimes \vec{S},$$

where $S_i = (\hbar/2)\sigma_i$ and σ_i are the Pauli matrices.

The operator that generates the corresponding dynamics is

$$K = \frac{1}{2m}\sum_i G_{p_i^2} \otimes \mathbb{1}/\hbar - \gamma\sum_i G_{B_i} \otimes S_i/\hbar.$$

As a simple model we take a magnetic field of the form $\vec{B} = (0, 0, B_3)$ with $B_3(\vec{q}) = \tilde{b}_0 - \tilde{b}_1 q_3$, and $\vec{q} = (q_1, q_2, q_3)$. We remark that this $\vec{B}$ cannot be an actual magnetic field, since $\nabla \cdot \vec{B} = -b_1 \neq 0$. We use it only as the simplest mathematical illustration of the types of behavior that can be expected. For a recent discussion of realistic models of the Stern-Gerlach experiments see [9–11].

Since the gradient of the magnetic field is only in the q_3 direction, we can restrict the model to one dimension. The Hilbert space is $\mathcal{K} = \mathcal{L}_2(\mathbb{R}^2, dp\, dq) \otimes \mathbb{C}^2$, where we denote $p \equiv p_3$, $q \equiv q_3$. Absorbing $-\gamma$ into the coefficients, the observable corresponding to the total energy can be written as

$$ H = \frac{1}{2m} p^2 \otimes \mathbb{1} + (b_0 + b_1 q) \otimes \sigma_3. $$

The corresponding Koopman-Schrödinger equation is

$$ i \frac{\partial \psi}{\partial t} = \left(\frac{1}{2m} G_{p^2} \otimes \mathbb{1} + (b_0 G_1 + b_1 G_q) \otimes \sigma_3 \right) \psi. $$

If we write $\psi(t) = (v_+(p, q, t), v_-(p, q, t))$ the above equations become two independent linear partial differential equations of first order:

$$ \frac{\partial v_+}{\partial t} = -\frac{p}{m} \frac{\partial v_+}{\partial q} + b_1 \frac{\partial v_+}{\partial p} - \frac{i}{\hbar} d(p, q) v_+, \tag{4.1} $$

$$ \frac{\partial v_-}{\partial t} = -\frac{p}{m} \frac{\partial v_-}{\partial q} - b_1 \frac{\partial v_-}{\partial p} + \frac{i}{\hbar} d(p, q) v_- \tag{4.2} $$

with

$$ d(p, q) := \left(b_0 + \frac{1}{2} b_1 q \right). $$

These equations can be solved explicitly, e.g. with the method of characteristics. The solution corresponding to an initial condition

$$ v_\pm(p, q, t = 0) = v_\pm^{(0)}(p, q) $$

is given by

$$ v_\pm(p, q, t) = e^{-i \delta_\pm (p, q, t)} v_\pm^{(0)} \left(p \pm b_1 t, \; q - \frac{p}{m} t \mp \frac{b_1}{2m} t^2 \right) \tag{4.3} $$

where $\delta_\pm(p, q, t)$ is a phase given by

$$ \hbar \delta_\pm = \pm \left[qt - \frac{p}{2m} t^2 \mp \frac{b_1}{6m} t^3 \right] \frac{b_1}{2} \pm b_0 t. $$

4.2. *Physical interpretation — classical-quantum entanglement*

The physical interpretation of $\psi = (v_+(p, q, t), v_-(p, q, t))$ is that $|v_\pm(p, q, t)|^2$ gives the probability for the center of mass of the particle to have a position q, a momentum p and a spin component $\pm\hbar/2$ in the q_3-direction. This dynamics can be interpreted as follows.

We consider initial conditions that are a tensor product of the spin and the momentum-position, i.e. of the form

$$\psi(t = 0) = \xi^{(0)}(p, q) \otimes \begin{pmatrix} s_+ \\ s_- \end{pmatrix},$$

with $|s_+|^2 + |s_-|^2 = 1$. Thus, in the initial state the spin and the center of mass are not entangled. We consider an initial state that is well-localized both in momentum p and in position q, described e.g. by narrow Gaussians

$$\xi^{(0)}(p, q) = \frac{1}{\sqrt{\pi w_p w_q}} e^{-\frac{(p-p_0)^2}{2w_p^2}} e^{-\frac{(q-q_0)^2}{2w_q^2}}$$

with $p_0 = 0$. The limit of complete localization corresponds to

$$\lim_{w_p, w_q \to 0} |\xi^{(0)}(p, q)|^2 = \delta(p - p_0)\delta(q - q_0).$$

We consider the following four examples of initial conditions:

$$\text{(i):} \quad \psi(t = 0) = \xi^{(0)}(p, q) \otimes \begin{pmatrix} 1 \\ 0 \end{pmatrix},$$

$$\text{(ii):} \quad \psi(t = 0) = \xi^{(0)}(p, q) \otimes \begin{pmatrix} 0 \\ 1 \end{pmatrix},$$

$$\text{(iii):} \quad \psi(t = 0) = \xi^{(0)}(p, q) \otimes \begin{pmatrix} 1 \\ 1 \end{pmatrix} \frac{1}{\sqrt{2}},$$

$$\text{(iv):} \quad \psi(t = 0) = \xi^{(0)}(p, q) \otimes \begin{pmatrix} s_+ \\ s_- \end{pmatrix}.$$

In (i) and (ii) the wave packet of the center of mass will follow a single trajectory without modification of its shape. It will move and accelerate in the direction $\pm q_3$, depending on the initial sign of the spin.

In cases (iii) and (iv) the state becomes a coherent superposition of two wave packets for the center of mass, that move in opposite directions. Both packets have the same shape, but the weight is given by a multiplicative factor $|s_\pm|^2$ determined by the initial state of the spin. We remark that in this case the spin and the center of mass are entangled, since the state $\psi(t)$

cannot be written as a tensor product of a spin state (s_1, s_2) and a center of mass state $\xi(p, q)$:

$$\psi(t) = v_+(p, q, t) \otimes \begin{pmatrix} 1 \\ 0 \end{pmatrix} + v_-(p, q, t) \otimes \begin{pmatrix} 0 \\ 1 \end{pmatrix} \neq \xi(p, q, t) \otimes \begin{pmatrix} s_1 \\ s_2 \end{pmatrix},$$

since this would imply $v_+(p, q, t) = \beta \, v_-(p, q, t)$, with a constant β.

This example shows that it is possible to entangle a classical and a quantum degree of freedom.

5. Berezin-Toeplitz Quantization — Geometric Quantization

In this section we review, with a minimum of mathematical formalism, some of the main ideas of Berezin-Toeplitz quantization and its relation with geometric quantization. We also summarize some elements of the theory of coherent states and their application to quantization.

In order to establish the relation between a quantum system and its classical counterpart we have to consider separately the relations between the Hilbert spaces, between the algebras of observables and between the derivations defining the dynamics.

The Berezin-Toeplitz quantization consists of selecting a subspace $\check{\mathcal{L}} \subset \mathcal{L}_K$ of the Koopman Hilbert space and a map that assigns to each multiplication operator M_f on $\mathcal{L}_K$ an operator T_{M_f} on $\check{\mathcal{L}}$ defined as the projection of M_f on $\check{\mathcal{L}}$. In general two operators $T_{M_{f_1}}, T_{M_{f_2}}$ corresponding to two different multiplication operators M_{f_1}, M_{f_2} do not commute with each other.

Geometric quantization can be viewed as an extension of this procedure to the quantization of the generators of the dynamics, i.e. the Hamiltonians. The link between the quantization of the observables and of the generators of the dynamics is given by the Tuynman relation [42].

5.1. *Selection of a polarization subspace*

As a first example we consider a system with one degree of freedom and phase space $\mathcal{M} = \mathbb{R}^2$. The Koopman Hilbert space is $\mathcal{L}_K = L_2(\mathbb{R}^2, dp\, dq)$, i.e. the square-integrable functions on the phase space.

The first step in the procedure of quantization is known in the literature on geometric quantization as the choice of a *"polarization"*, which here we formulate as the choice of a *polarization subspace* $\check{\mathcal{L}} \subset \mathcal{L}_K$. We present a simple description of the construction in terms of action-angle variables

I, θ, defined by the canonical transformation

$$I = \frac{1}{2}(p^2/\beta_0 + q^2\beta_0), \qquad\qquad p = \sqrt{2I\beta_0}\sin\theta, \qquad (5.1)$$

$$\theta = \arctan(p/\sqrt{\beta_0}, q\sqrt{\beta_0}), \qquad q = \sqrt{2I/\beta_0}\cos\theta, \qquad (5.2)$$

where β_0 is an arbitrary reference constant (with units of a mass times a frequency $\beta_0 = m_0\omega_0$, or equivalently of an action times the square of a length). The function $\arctan(y,x)$ of two arguments is defined as the single-valued function that gives the unique angle $\theta \in [0, 2\pi]$ such that $\cos\theta = y/\sqrt{y^2 + x^2}$ and $\sin\theta = x/\sqrt{y^2 + x^2}$.

5.1.1. Construction of a basis of $\mathcal{L}_K$

We will use the following basis of $L_2(\mathbb{R}^2, dp\, dq)$, expressed in action-angle coordinates:

$$\tilde{\xi}_{m',m}(I, \theta) := \nu_{m',m}\, e^{-\frac{1}{2\lambda}I}I^{(m'+m)/2}e^{i(m-m')\theta}, \quad m, m' \in \mathbb{N}_0 = \{0, 1, 2, \ldots\} \tag{5.3}$$

where $\nu_{m',m}$ is the normalization factor, and λ is an arbitrary fixed real constant. Since the argument of the exponential should be dimensionless, λ must have the dimension of an action. We remark that a constant that makes the variable dimensionless in the function $I^{(m'+m)/2}$ is included in the normalization factor $\nu_{m',m}$, to simplify the notation.

One can verify that (5.3) is a basis of $L_2(\mathbb{R}^2, dp\, dq) = L_2(\mathbb{R}, dp) \otimes L_2(\mathbb{R}, dq)$ by first considering the known basis of $L_2(\mathbb{R}, dp)$

$$H_{m'}(p/\sqrt{2\lambda\beta_0})e^{-\frac{1}{2\lambda\beta_0}p^2/2}, \qquad m' \in \mathbb{N}_0$$

where $H_{m'}$ are the Hermite polynomials, and the basis of $L_2(\mathbb{R}, dq)$

$$H_m(q\sqrt{\beta_0/(2\lambda)})e^{-\frac{\beta_0}{2\lambda}q^2/2}, \qquad m \in \mathbb{N}_0.$$

The set of functions

$$p^{m'}q^m e^{-\frac{1}{2\lambda}\left(p^2/\beta_0 + q^2\beta_0\right)/2}, \qquad m, m' \in \mathbb{N}_0$$

is therefore a basis $L_2(\mathbb{R}^2, dp\, dq)$, and defining the dimensionless complex variable

$$z := \frac{1}{\sqrt{2\lambda}}(q\sqrt{\beta_0} + ip/\sqrt{\beta_0}) =: \sqrt{\frac{I}{\lambda}}e^{i\theta}, \tag{5.4}$$

another basis is given by the functions

$$\tilde{\xi}_{m',m} = \tilde{\nu}_{m',m}\, e^{-\frac{1}{2}|z|^2}\, z^{*m'} z^m \equiv \nu_{m',m}\, e^{-\frac{1}{2\lambda}I}I^{\frac{m'+m}{2}}e^{i(m-m')\theta},$$

which coincides with (5.3). We remark that the measure $d\mu$ expressed in the complex variables z is

$$d\mu(z) := dq\, dp = 2\lambda d^2 z, \quad \text{with} \quad d^2 z = dz_r\, dz_i, \quad z = z_r + iz_i. \quad (5.5)$$

We can relabel the basis vectors through

$$k := m - m', \qquad m = \frac{1}{2}(n + k), \qquad (5.6)$$

$$n := m + m', \qquad m' = \frac{1}{2}(n - k), \qquad (5.7)$$

and write the basis as

$$\xi_{n,k}(I, \theta) := \nu_{n,k} e^{-\frac{1}{2\lambda} I} I^{n/2} e^{ik\theta}, \qquad (5.8)$$

with $n \in \mathbb{N}_0$, $k \in \{-n, -n+2, -n+4, \ldots, n-2, n\}$, $\nu_n = (n!\, 2\pi\lambda^{n+1})^{-1/2}$.

5.1.2. *Selection of a polarization subspace $\check{\mathcal{L}} \subset \mathcal{L}_K$ by the choice of a subset of the basis*

The selection of the polarization subspace can be performed by selecting a subset of this basis. In the standard Berezin quantization one selects the subspace $\check{\mathcal{L}} \subset \mathcal{L}_K$ as follows:

First one chooses a particular value for the constant $\lambda = \hbar$, where $\hbar$ is equal to the constant that we had introduced in Eq. (3.3), when we introduced a phase in the unitary Koopman evolution. Then one selects the subspace generated by the subset $\{\eta_n\} \subset \{\xi_{n,k}\}$ of the basis functions defined as

$$\eta_n(I, \theta) := \xi_{n,k=-n}(I, \theta) \equiv \nu_n\, e^{-\frac{1}{2\hbar} I} I^{n/2} e^{-in\theta}, \qquad n \in \mathbb{N}_0 \quad (5.9)$$

$$\equiv \tilde{\nu}_n\, z^{*n} e^{-zz^*/2}, \qquad (5.10)$$

with $\nu_n = (n!\, 2\pi\, \hbar^{n+1})^{-1/2}$ and $\tilde{\nu}_n = (n!\, 2\pi\, \hbar)^{-1/2}$.

We remark that different choices of the constant λ lead to different subspaces, e.g. the function $e^{-\frac{1}{2\lambda'} I}$ is not contained in the subspace $\check{\mathcal{L}}$ if $\lambda' \neq \hbar$. Thus the choice of λ equal to Planck's constant $\hbar$ is non-trivial, in the sense that it is not just a conventional choice of units, but it is an essential ingredient in the definition of the quantum model. Its numerical value in any given system of units must be determined by comparison with experiments.

5.1.3. Definition of an isomorphism between $\check{\mathcal{L}} \subset \mathcal{L}_K$ and Fock space

The basis (5.9) of $\check{\mathcal{L}}$ is labeled by a single index $n \in \mathbb{N}_0$. One can define an isomorphism Ξ between the subspace $\check{\mathcal{L}}$ (whose elements are functions of p, q) and a Hilbert space $\mathcal{H}$, that will be the Hilbert space of the quantum system, which can be defined formally as the space generated by a set of orthonormal states $\{|n\rangle\}_{n \in \mathbb{N}_0}$. The isomorphism is defined by

$$\Xi : \eta_n \mapsto |n\rangle$$

which in the Dirac notation can be written as

$$\Xi := \sum_n |n\rangle\langle\eta_n|.$$

As concrete examples for the quantum Hilbert space $\mathcal{H}$ we consider two examples:

(i) We can take $\mathcal{H}$ as the abstract Fock space $\mathcal{F}$ constructed from a ground state $|0\rangle$ and the creation operator $a^\dagger : |n\rangle := \nu_n (a^\dagger)^n |0\rangle$.

(ii) One can take $\mathcal{H} = L_2(\mathbb{R}, dx)$ and for $|n\rangle$ the basis of eigenfunctions of the Hamiltonian of a harmonic oscillator

$$H_{h.o.} := -\frac{\hbar^2}{2m}\frac{d^2}{dx^2} + \frac{m\omega^2}{2}x^2,$$

with m and ω such that $m\omega = \beta_0$, where β_0 is the constant used in (5.1), i.e.

$$\Xi : \eta_n \mapsto |n\rangle = \varphi_n(x) = \nu_n H_n(x\sqrt{m\omega/\hbar})e^{-\frac{1}{2}x^2 m\omega/\hbar}, \qquad (5.11)$$

where H_n are the Hermite polynomials.

Remark. Since $z^* := \sqrt{\frac{I}{\hbar}}e^{-i\theta}$, by Eq. (5.8) the subspace $\check{\mathcal{L}}$ can be also identified as being isomorphic to the Hilbert space of anti-holomorphic functions $g(z^*)$ [30–32], with scalar product $\langle g_1, g_2 \rangle := \int_{\mathcal{M}} dz\, e^{-|z|^2} g_1^*(z^*)\, g_2(z^*)$, which is the usual formulation in the literature on geometric and Berezin quantization.

Summary. The polarization subspace $\check{\mathcal{L}} \subset \mathcal{L}_K$ is the subspace generated by the orthonormal set of functions

$$\eta_n := \nu_n\, I^{n/2}e^{-in\theta}e^{-\frac{1}{2\hbar}I} \equiv \tilde{\nu}_n\, z^{*n}e^{-zz^*/2}, \qquad n \in \mathbb{N}_0, \qquad (5.12)$$

with $\nu_n = (n!\, 2\pi\, \hbar^{n+1})^{-1/2}$ and $\tilde{\nu}_n = (n!\, 2\pi\, \hbar)^{-1/2}$. The isomorphism $\Xi : \eta_n \mapsto |n\rangle$ gives the representation in the quantum Hilbert space $\mathcal{H}$.

5.2. *Toeplitz quantization of the observables*

The Toeplitz quantization of operators of the classical Hilbert space consists simply of taking the projection of the operator on the polarization subspace: To a multiplication operator M_f acting on $\mathcal{L}_K$ one associates an operator T_{M_f}

$$M_f \mapsto T_{M_f} := P_{\check{\mathcal{L}}} M_f P_{\check{\mathcal{L}}}, \tag{5.13}$$

where $P_{\check{\mathcal{L}}}$ is the orthogonal projection from $\mathcal{L}_K$ to the polarization subspace $\check{\mathcal{L}}$. By composition with the isomorphism Ξ one defines the associated operator on $\mathcal{H}$:

$$M_f \mapsto \widehat{f} \equiv \widehat{T}_f := \Xi \, T_f \, \Xi^{-1} = \Xi \, P_{\check{\mathcal{L}}} M_f P_{\check{\mathcal{L}}} \, \Xi^{-1},$$

which can be expressed in terms of the bases $\xi_{n,k=-n}$ of $\check{\mathcal{L}}$ and $|n\rangle$ of $\mathcal{H}$ as

$$M_f \mapsto \widehat{f} \equiv \widehat{T}_f := \sum_{n',n} |n'\rangle \langle \eta_{n'} | M_f | \eta_n \rangle \langle n|. \tag{5.14}$$

One can calculate the matrix elements for some of the basic polynomial functions explicitly:

$$\langle \eta_{n'} | z | \eta_n \rangle = \delta_{n',n-1} \sqrt{n}\,, \tag{5.15}$$

$$\langle \eta_{n'} | z^* | \eta_n \rangle = \delta_{n',n+1} \sqrt{n+1} \tag{5.16}$$

and

$$\langle \eta_{n'} | z^m (z^*)^k | \eta_n \rangle = \delta_{n',n-m+k}$$
$$\times \sqrt{(n+k-m+1)\cdots(n+k-1)(n+k)\,(n+k)(n+k-1)\cdots(n+1)},$$

which leads to their identification in terms of creation-annihilation operators $a^\dagger, a$ (either in the abstract Fock space $\mathcal{F}$ or in $L_2(\mathbb{R}, dx)$):

$$\widehat{z} = \hat{a}\,, \tag{5.17}$$

$$\widehat{z^*} = \hat{a}^\dagger\,, \tag{5.18}$$

$$\widehat{T}_{z^k z^{*m}} \equiv \widehat{z^k z^{*m}} = \hat{a}^k \, \hat{a}^{\dagger m}. \tag{5.19}$$

We remark that the Berezin-Toeplitz quantization with the chosen polarization subspace yields the operators (5.19) in anti-normal ordering, i.e. with all the $\hat{a}^\dagger$ on the right.

If we use the representation $\mathcal{H} = L_2(\mathbb{R}, dx)$ defined by the isomorphism (5.11), and

$$q = \sqrt{\frac{\hbar}{2\beta_0}}(z + z^*), \qquad p = \sqrt{\frac{\hbar\beta_0}{2}}(z - z^*), \tag{5.20}$$

$$\hat{x} = \sqrt{\frac{\hbar}{2\beta_0}}(\hat{a} + \hat{a}^\dagger), \qquad \hat{p} = \sqrt{\frac{\hbar\beta_0}{2}}(\hat{a} + \hat{a}^\dagger), \tag{5.21}$$

we obtain

$$\hat{q} = M_x =: \hat{x} \qquad \text{(multiplication by } x\text{)}, \tag{5.22}$$

$$\hat{p} = -i\hbar\frac{\partial}{\partial x} =: \hat{p}, \tag{5.23}$$

$$\widehat{p^2} = (\hat{p})^2 + \frac{\hbar\beta_0}{2}\mathbb{1} = -\hbar^2\frac{\partial^2}{\partial x^2} + \frac{\hbar\beta_0}{2}\mathbb{1}, \tag{5.24}$$

$$\widehat{q^2} = (\hat{q})^2 + \frac{\hbar}{2\beta_0}\mathbb{1} = \hat{x}^2 + \frac{\hbar}{2\beta_0}\mathbb{1}, \tag{5.25}$$

$$\widehat{I} = \frac{1}{2}\left(\frac{1}{\beta_0}\hat{p}^2 + \beta_0\hat{x}^2\right) = -i\hbar\frac{\partial}{\partial\theta}, \tag{5.26}$$

$$\widehat{H_{h.o.}} = \omega\widehat{I} = -\frac{\hbar^2}{2m}\frac{\partial^2}{\partial x^2} + \frac{m\omega^2}{2}\hat{x}^2 + \frac{\hbar\omega}{2}, \tag{5.27}$$

where $H_{h.o.} = \omega I = \omega\hbar z z^* = \frac{1}{2m}p^2 + \frac{m\omega^2}{2}q^2$.

5.3. Toeplitz quantization of the generators of the dynamics — geometric quantization

The Toeplitz quantization, that we first have defined for multiplication operators as the projection into the polarization subspace $\check{\mathcal{L}}$, can be extended to the differential operators of the generators of the dynamics:

$$G_f \mapsto T_{G_f} := P_{\check{\mathcal{L}}}G_f P_{\check{\mathcal{L}}}. \tag{5.28}$$

By composition with the isomorphism Ξ one defines the associated operator on the quantum Hilbert space $\mathcal{H}$:

$$G_g \mapsto \widehat{G_g} := \Xi\, P_{\check{\mathcal{L}}}G_g P_{\check{\mathcal{L}}}\, \Xi^{-1},$$

which can be expressed in terms of the bases η_n of $\check{\mathcal{L}}$ and $|n\rangle$ of $\mathcal{H}$ as

$$G_g \mapsto \widehat{G_g} = \sum_{n',n}|n'\rangle\langle\eta_{n'}|G_g|\eta_n\rangle\langle n|. \tag{5.29}$$

The Schrödinger equation in the Hilbert space $\mathcal{H}$, corresponding to a classical Hamilton function H_{cl} is thus given by

$$i\frac{\partial\psi}{\partial t} = \widehat{G}_{H_{cl}}\psi, \quad \text{i.e.} \quad i\hbar\frac{\partial\psi}{\partial t} = \hat{H}_{H_{cl}}\psi, \quad \text{with} \quad \hat{H}_{H_{cl}} := \hbar\widehat{G}_{H_{cl}}.$$

The Poisson brackets can be expressed in terms of the complex coordinates (5.4), choosing $\lambda = \hbar$, as

$$\{h, f\} := \frac{i}{\hbar}\sum_j \frac{\partial h}{\partial z_j}\frac{\partial f}{\partial z_j^*} - \frac{\partial h}{\partial z_j^*}\frac{\partial f}{\partial z_j},$$

and Λ_f as

$$\Lambda_f = -\frac{1}{2}\sum_j z\frac{\partial f}{\partial z} + z^*\frac{\partial f}{\partial z^*}.$$

For a Hamilton function of the form $f = z^k z^{*m}$ the quantized generator of the dynamics is given by

$$\hbar\widehat{G}_f = \hat{a}^k\hat{a}^{\dagger m} - km\,\hat{a}^{(k-1)}\hat{a}^{\dagger(m-1)}. \tag{5.30}$$

This result can be obtained by the following steps: We first determine

$$\Lambda_f = -\frac{k+m}{2}z^k z^{*m}, \tag{5.31}$$

$$\hbar\widetilde{X}_f = kz^{(k-1)}z^{*m}\frac{\partial}{\partial z^*} - mz^k z^{*(m-1)}\frac{\partial}{\partial z}, \tag{5.32}$$

$$\hbar G_f = f + \Lambda_f + \hbar\tilde{X}_f \tag{5.33}$$

$$= \left(1 - \frac{k+m}{2}\right)z^k z^{*m} + kz^{(k-1)}z^{*m}\frac{\partial}{\partial z^*} - mz^k z^{*(m-1)}\frac{\partial}{\partial z}. \tag{5.34}$$

Then we determine $\widehat{X}_f := \Xi\,P_{\tilde{\mathcal{L}}}\tilde{X}_f P_{\tilde{\mathcal{L}}}\,\Xi^{-1}$, as

$$\hbar\widehat{X}_f = \frac{k+m}{2}\hat{a}^k\hat{a}^{\dagger m} - km\,\hat{a}^{(k-1)}\hat{a}^{\dagger(m-1)}, \tag{5.35}$$

which combined with

$$\widehat{T}_{z^k z^{*m}} = \hat{a}^k\hat{a}^{\dagger m} \tag{5.36}$$

yields (5.30).

Remark. Eq. (5.30) is a special case of Tuynman's relation [42], which allows to express the quantization of the generator of the dynamics G_f in terms of the Toeplitz quantization of an associated observable $\tau(f)$:

Proposition (Tuynman's relation [42]).

$$\hbar \widehat{G}_f = \widehat{T}_{\tau(f)}, \quad \text{with} \quad \tau(f) := f - \frac{\partial^2 f}{\partial z \partial z^*}. \tag{5.37}$$

This relation can be written in a more general context as $\tau(f) := f + \frac{\hbar}{4}\Delta_{dR}$, where Δ_{dR} is the de Rham Laplacian, which in our case is given by $\Delta_{dR} = -\frac{4}{\hbar}\frac{\partial^2}{\partial z \partial z^*}$. We notice that in [42] the complex variables z are defined with a convention that differs from ours by a factor $\sqrt{2}$. We remark that

$$\frac{\partial^2 f}{\partial z \partial z^*} = \frac{\hbar}{2}\left(\frac{1}{\beta_0}\frac{\partial^2 f}{\partial q^2} + \beta_0 \frac{\partial^2 f}{\partial p^2} \right). \tag{5.38}$$

In Table 5.1 we give for some examples of the expressions of the operators $\Lambda_f, \widetilde{X}_f, \widehat{T}_f, \widehat{G}_f$. We use the notation $\hat{x} = M_x; \quad \hat{p} = -i\hbar\frac{\partial}{\partial x}$. In Table 5.1 we have expressed the quantized operator corresponding to the action I, in the phase representation [12–14] where the Hilbert space $\mathcal{H}$ is generated by the functions $\{e^{in\theta}, \quad n \in \mathbb{N}_0\}$. We remark that there is no simple explicit expression for the $\widehat{T}_V, \widehat{G}_V$ corresponding to a general potential $V(q)$.

For the harmonic oscillator we have:

$$H_{h.o.} = \omega\left(\frac{1}{2m\omega}p^2 + \frac{1}{2}m\omega q^2 \right) = \omega I = \hbar\omega z z^*, \tag{5.39}$$

$$\Lambda_{H_{h.o.}} = -H_{h.o.}, \tag{5.40}$$

$$\widetilde{X}_{H_{h.o.}} = -i\omega\left(\frac{1}{2m\omega}p\frac{\partial}{\partial q} + \frac{1}{2}m\omega q\frac{\partial}{\partial p} \right) = -i\omega\frac{\partial}{\partial\theta}, \tag{5.41}$$

$$G_{H_{h.o.}} = \widetilde{X}_{H_{h.o.}}, \tag{5.42}$$

$$\widehat{T}_{H_{h.o.}} = \omega\left(\frac{1}{2m\omega}\hat{p}^2 + \frac{1}{2}m\omega\hat{x} \right) + \frac{\hbar\omega}{2} \tag{5.43}$$

$$= \hbar\omega\hat{a}\hat{a}^\dagger = \hbar\omega\left(\hat{a}^\dagger\hat{a} + \frac{1}{2} \right) + \frac{\hbar\omega}{2}, \tag{5.44}$$

$$\widehat{H}_{h.o.} \equiv \hbar\widehat{G}_{H_{h.o.}} = \hbar\widehat{X}_{H_{h.o.}} \tag{5.45}$$

$$= \hbar\omega(\hat{a}\hat{a}^\dagger - 1) = \hbar\omega\hat{a}^\dagger\hat{a} \tag{5.46}$$

$$= \left(\frac{1}{2m}\hat{p}^2 + \frac{1}{2}m\omega^2\hat{x} \right) - \frac{\hbar\omega}{2}. \tag{5.47}$$

5.4. *Quantization by coherent states*

An alternative formulation of the quantization of multiplication operators on $\mathcal{L}_K$ using coherent states was proposed in [33–36]. As we will discuss

Table 5.1. Some examples of the expressions of the operators Λ_f, $\tilde{X}_f$, $\widehat{T}_f$, $\widehat{G}_f$.

f	Λ_f	$\tilde{X}_f$	G_f	$\widehat{T}_f$	$\hbar\widehat{G}_f = \widehat{H}_f$
1	0	0	$\frac{1}{\hbar}\mathbb{1}$	1	1
$z^k z^{*m}$	$-\frac{k+m}{2}z^k z^{*m}$	(5.32)	(5.34)	$\hat{a}^k\hat{a}^{\dagger m}$	$\hat{a}^k\hat{a}^{\dagger m} - km\hat{a}^{(k-1)}\hat{a}^{\dagger(m-1)}$
q	$-\frac{1}{2}q$	$i\frac{\partial}{\partial p}$	$\frac{1}{2\hbar}q + i\frac{\partial}{\partial p}$	$\hat{x}$	$\hat{x}$
p	$-\frac{1}{2}p$	$-i\frac{\partial}{\partial q}$	$\frac{1}{2\hbar}p - i\frac{\partial}{\partial q}$	$\hat{p}$	$\hat{p}$
q^2	$-q^2$	$i2q\frac{\partial}{\partial p}$	$i2q\frac{\partial}{\partial p}$	$\hat{x}^2 + \frac{\hbar}{2\beta_0}$	$\hat{x}^2 - \frac{\hbar}{2\beta_0}$
p^2	$-p^2$	$-i2p\frac{\partial}{\partial q}$	$-i2p\frac{\partial}{\partial q}$	$\hat{p}^2 + \frac{\hbar\beta_0}{2}$	$\hat{p}^2 - \frac{\hbar\beta_0}{2}$
I	$-I$	$-i\{I, \}$	$-i\{I, \} = -i\frac{\partial}{\partial\theta}$	$-i\hbar\frac{\partial}{\partial\theta} + \hbar\mathbb{1}$	$-i\hbar\frac{\partial}{\partial\theta}$
$V(q)$	$-\frac{1}{2}q\frac{\partial V}{\partial q}$	$i\frac{\partial V}{\partial q}\frac{\partial}{\partial p}$	$\frac{1}{\hbar}V - \frac{\partial V}{\partial q}\left(\frac{q}{2\hbar} - i\frac{\partial}{\partial p}\right)$		

below, this formulation yields the same quantized operators as the Toeplitz quantization. It can also be extended to yield the same quantization of the generators of the dynamics.

5.4.1. *Definition of coherent states*

There are several definitions of coherent states that emphasize different types of properties [15, 17, 27]: Minimization of the Heisenberg uncertainty relations [25], group theoretical properties [20, 23], annihilation operator eigenfunctions [26].

For the purpose of establishing relations between classical and quantum systems one can use a definition that addresses only one property that is shared by all the other definitions: We consider a Hilbert space $\mathcal{H}$ and a phase space $\mathcal{M}$ with a volume measure $d\mu$.

Definition. A *complete set of vectors indexed by the points of $\mathcal{M}$*, $\{|\zeta_{z_0}^{\mathcal{H}}\rangle \in \mathcal{H}\}_{z_0 \in \mathcal{M}}$, is defined by a continuous map $\mathcal{M} \to \mathcal{H}$, $z_0 \mapsto |\zeta_{z_0}\rangle$, such that

$$\int_{\mathcal{M}} d\mu(z_0)\, |\zeta_{z_0}^{\mathcal{H}}\rangle\langle\zeta_{z_0}^{\mathcal{H}}| = \mathbb{1}_{\mathcal{H}}. \tag{5.48}$$

Definition. The corresponding *coherent states* are defined as the normalized vectors

$$|C_{z_0}^{\mathcal{H}}\rangle := \frac{1}{\nu(z_0)}|\zeta_{z_0}^{\mathcal{H}}\rangle, \tag{5.49}$$

with $\nu^2(z_0) := \langle\zeta_{z_0}^{\mathcal{H}}|\zeta_{z_0}^{\mathcal{H}}\rangle$. The non-normalized vectors $|\zeta_{z_0}^{\mathcal{H}}\rangle$, will also be called *unnormalized coherent states*.

5.4.2. *Construction of the coherent states determined by the selection of a polarization subspace*

In [27, 33, 36] a general construction of coherent states was proposed which is based on the selection of a polarization subspace $\check{\mathcal{L}}$. We assume that the elements of $\check{\mathcal{L}}$ are continuous functions. One defines for each point $z_0 \in \mathcal{M}$ of the phase space an *evaluation functional*, that assigns to each function its value at the point z_0:

$$\delta_{z_0} : \check{\mathcal{L}} \to \mathbb{C} \tag{5.50}$$

$$|\xi\rangle \mapsto \xi(z_0). \tag{5.51}$$

Under the assumption that this linear map is continuous, by Riesz's theorem [39] there is a unique vector $|\zeta_{\underline{z}_0}\rangle \in \mathcal{H}$ such that $\forall |\xi\rangle \in \check{\mathcal{L}}$

$$\delta_{\underline{z}_0}|\xi\rangle = \langle \zeta_{\underline{z}_0} \mid \xi \rangle.$$

One can give an explicit expression of the functional $\langle \zeta_{\underline{z}_0}|$ in terms of the arbitrary orthonormal basis of continuous functions $\{|\eta_n\rangle\} \in \check{\mathcal{L}}$:

$$\delta_{\underline{z}_0} \equiv \langle \zeta_{\underline{z}_0}| = \sum_n \eta_n(\underline{z}_0)\langle \eta_n|, \tag{5.52}$$

since, $\forall |\xi\rangle \in \check{\mathcal{L}}$, $|\xi\rangle = \sum_n |\eta_n\rangle\langle\eta_n|\xi\rangle$ and thus,

$$\delta_{\underline{z}_0}|\xi\rangle = \sum_n \delta_{\underline{z}_0}(|\eta_n\rangle)\langle\eta_n|\xi\rangle = \sum_n \eta_n(\underline{z}_0)\langle\eta_n|\xi\rangle.$$

The corresponding vector can thus be written as

$$|\zeta_{\underline{z}_0}\rangle = \sum_n \eta_n^*(\underline{z}_0)|\eta_n\rangle. \tag{5.53}$$

The set of vectors $|\zeta_{\underline{z}_0}\rangle$ satisfies the following completeness relation:

$$\int_{\mathcal{M}} d\mu(\underline{z}_0) \; |\zeta_{\underline{z}_0}\rangle\langle\zeta_{\underline{z}_0}| = \mathbb{1}_{\check{\mathcal{L}}}, \tag{5.54}$$

since

$$\int_{\mathcal{M}} d\mu(\underline{z}_0)|\zeta_{\underline{z}_0}\rangle\langle\zeta_{\underline{z}_0}| = \sum_{n',n}\left(\int_{\mathcal{M}} d\mu(\underline{z}_0) \; \eta_{n'}^*(\underline{z}_0)\eta_n(\underline{z}_0)\right)|\eta_{n'}\rangle\langle\eta_n|$$

$$= \sum_{n',n} \delta_{n',n}|\eta_{n'}\rangle\langle\eta_n| = \mathbb{1}_{\check{\mathcal{L}}},$$

where we have used $\int_{\mathcal{M}} d\mu(\underline{z}_0) \; \eta_{n'}^*(\underline{z}_0)\eta_n(\underline{z}_0) = \delta_{n',n}$.

We remark that, since the evaluation vector $|\zeta_{\underline{z}_0}\rangle$ is a function of $\underline{z}$ which belongs to the subspace $\check{\mathcal{L}}$, we can write it as

$$\zeta_{\underline{z}_0}(\underline{z}) = \sum_n \eta_n^*(\underline{z}_0) \; \eta_n(\underline{z}).$$

In the limit case when the polarization subspace coincides with the total Hilbert space $\check{\mathcal{L}} = \mathcal{L}_K$, the evaluation vector tends formally to a Dirac delta function:

$$\zeta_{\underline{z}_0}(\underline{z}) \xrightarrow[\mathcal{H} \to L_K]{} \delta(\underline{z} - \underline{z}_0).$$

The *coherent states* $|C_{\underline{z}_0}\rangle$ determined by the choice of the subspace $\check{\mathcal{L}}$ are defined by normalizing the vectors $|\zeta_{\underline{z}_0}\rangle$:

$$|C_{\underline{z}_0}\rangle := \frac{1}{\nu(\underline{z}_0)}|\zeta_{\underline{z}_0}\rangle,$$

with

$$\nu^2(\underline{z}_0) := \langle\zeta_{\underline{z}_0}|\zeta_{\underline{z}_0}\rangle = \sum_{n',n} \eta_{n'}(\underline{z}_0)\eta_n^*(\underline{z}_0)\langle\eta_{n'}|\eta_n\rangle = \sum_n |\eta_n(\underline{z}_0)|^2.$$

One can define the analogue of the the evaluation vectors and of the evaluation functionals in the quantum Hilbert space $\mathcal{H}$ (e.g. in Fock space $\mathcal{F}$ or in $L_2(\mathbb{R},\ dx)$) by

$$|\zeta_{\underline{z}_0}^{\mathcal{H}}\rangle := \Xi\big(|\zeta_{\underline{z}_0}\rangle\big) = \sum_n \eta_n^*(\underline{z}_0)|n\rangle \qquad (5.55)$$

$$\delta_{\underline{z}_0}^{\mathcal{H}} \equiv \langle\zeta_{\underline{z}_0}^{\mathcal{H}}| := \sum_n \eta_n(\underline{z}_0)\langle n|, \qquad (5.56)$$

where $\{|n\rangle\}$ is the orthogonal basis of the space $\mathcal{H}$.

Remark. As we will see in Section 5.4.3 in the quantization of observables and of the dynamics we don't actually use the normalized coherent states $|C_{\underline{z}_0}\rangle$, but directly the *evaluation functionals* $\langle\zeta_{\underline{z}_0}|$, $\langle\zeta_{\underline{z}_0}^{\mathcal{H}}|$ and their duals, the *evaluation vectors* $|\zeta_{\underline{z}_0}\rangle$, $|\zeta_{\underline{z}_0}^{\mathcal{H}}\rangle$ (that we will also call *unnormalized coherent states*). The essential property are the completeness relations (5.48), (5.54). We remark, however, that for the standard Glauber coherent states [18, 19], as well as for the spin or atomic coherent states of Gilmore [22] and of Perelomov [21], the normalization factor $\nu(\underline{z}_0)$ is independent of $\underline{z}_0$, and thus the coherent states and the unnormalized coherent states are related by just a multiplicative constant.

In summary, the choice of a polarization subspace $\check{\mathcal{L}}$ and the isomorphism Ξ define the coherent states (5.56) uniquely. We will see in Section 6.0.2 an inverse property: A given set of coherent states in $\mathcal{H}$ determines uniquely a polarization subspace $\check{\mathcal{L}} \subset \mathcal{L}_K$ and the isomorphism Ξ.

5.4.3. *Coherent state quantization of observables*

Using these unnormalized coherent states one can associate to each multiplication operator M_f on $\mathcal{L}_K$ an operator C_{M_f} on the subspace $\check{\mathcal{L}}$:

$$C_{M_f} := \int_{\mathcal{M}} d\mu(\underline{z}_0)f(\underline{z}_0)|\zeta_{\underline{z}_0}\rangle\langle\zeta_{\underline{z}_0}|. \qquad (5.57)$$

The following argument shows that the quantized operator C_{M_f} defined by Eq. (5.57) with the coherent states, is identical to the Toeplitz operator (5.13) defined by projection on the polarization subspace $\check{\mathcal{L}}$: $C_{M_f} \equiv T_{M_f}$.

By inserting the representations (5.52), (5.53) into (5.57), we obtain

$$C_{M_f} = \sum_{n',n} \int_{\mathcal{M}} d\mu(\underline{z}_0)\ f(\underline{z}_0)\ \eta_{n'}^*(\underline{z}_0)\ \eta_n(\underline{z}_0)\ |\eta_{n'}\rangle\langle\eta_n|$$

$$= \sum_{n',n} |\eta_{n'}\rangle \left(\int_{\mathcal{M}} d\mu(\underline{z}_0)\ f(\underline{z}_0)\ \eta_{n'}^*(\underline{z}_0)\ \eta_n(\underline{z}_0) \right) \langle\eta_n|,$$

and using the fact that

$$\int_{\mathcal{M}} d\mu(\underline{z})\ f(\underline{z})\ \eta_{n'}^*(\underline{z})\ \eta_n(\underline{z}) \equiv \langle\eta_{n'}|M_f|\eta_n\rangle$$

and defining the projector $P_{\check{\mathcal{L}}} := \sum_n |\eta_n\rangle\langle\eta_n|$ into the polarization subspace $\check{\mathcal{L}}$, we can write

$$C_{M_f} = \sum_{n',n} |\eta_{n'}\rangle\langle\eta_{n'}|M_f|\eta_n\rangle\langle\eta_n|$$

$$= P_{\check{\mathcal{L}}} M_f P_{\check{\mathcal{L}}} \equiv T_{M_f},$$

which is the Toeplitz operator (5.13).

Using the analogues of the evaluation vectors and functionals defined by Eqs. (5.55) in the quantum Hilbert space $\mathcal{H}$ (i.e. on Fock space $\mathcal{F}$ or in $L_2(\mathbb{R}, dx)$), the quantized operator $\hat{f}$ corresponding to the observable f is given by

$$\hat{f} = \int_{\mathcal{M}} d\mu(\underline{z}_0)\ f(\underline{z}_0)\ |\zeta_{\underline{z}_0}^{\mathcal{H}}\rangle\langle\zeta_{\underline{z}_0}^{\mathcal{H}}|. \tag{5.58}$$

5.4.4. *Coherent state quantization of the generators of the dynamics*

The Toeplitz quantization of the generator of the dynamics can also be expressed in terms of coherent states. The formula (5.57), originally defined for multiplication operators, can be extended to the differential operators X_g appearing in the generators G_g

$$C_{X_g} := \int_{\mathcal{M}} d\mu(\underline{z}_0)\ |\zeta_{\underline{z}_0}\rangle\ X_g^{\underline{z}_0}(\langle\langle\zeta_{\underline{z}_0}|) \tag{5.59}$$

where the notation $X_g^{z_0}$ indicates that the differential operator acts on the variables z_0 and not on z:

$$X_g^{z_0}(\langle\zeta_{z_0}|) = X_g^{z_0}\sum_n \eta_n(z_0)\langle\eta_n| = \sum_n X_g(\eta_n(z_0))\langle\eta_n|$$

$$= \sum_n \{g(z_0), \eta_n(z_0)\}\langle\eta_n|.$$

The coherent state quantization of the generators of the dynamics can thus be written as

$$C_{G_g} := \int_{\mathcal{M}} d\mu(z_0)\, |\zeta_{z_0}\rangle\, G_g^{z_0}(\langle\zeta_{z_0}|). \tag{5.60}$$

The result is again the same one as the one obtained by Toeplitz quantization:

$$C_{G_g} = T_{G_g}.$$

This can ve verified by an argument along the same line as the one for the multiplication operators:

$$C_{G_g} = \int_{\mathcal{M}} d\mu(z_0)\, |\zeta_{z_0}\rangle\, G_g^{z_0}(\langle\zeta_{z_0}|) \tag{5.61}$$

$$= \sum_{n',n} \int_{\mathcal{M}} d\mu(z_0)\, \eta_{n'}^*(z_0)\, G_g^{z_0}(\eta_n(z_0))\, |\eta_{n'}\rangle\langle\eta_n| \tag{5.62}$$

$$= \sum_{n',n} |\eta_{n'}\rangle\left(\int_{\mathcal{M}} d\mu(z_0)\, \eta_{n'}^*(z_0)\, G_g^{z_0}(\eta_n(z_0))\right)\langle\eta_n| \tag{5.63}$$

$$= \sum_{n',n} |\eta_{n'}\rangle\langle\eta_{n'}|\, G_g\, |\eta_n\rangle\langle\eta_n| \tag{5.64}$$

$$= P_{\check{\mathcal{L}}}G_g P_{\check{\mathcal{L}}} \equiv T_{G_g}. \tag{5.65}$$

6. Dequantization by Coherent States

Coherent states can be used for the opposite process, called *dequantization*, which is the construction of a classical system for a given quantum model [37, 40].

The general problem of dequantization can be formulated as follows: Given a quantum system defined on a Hilbert space $\mathcal{H}$, with observables $\hat{A}$ and a dynamics generated by a Hamiltonian $\hat{H}$, the goal is to find

(a) a phase space manifold $\mathcal{M}$ and a measure $d\mu$,
(b) a subspace $\check{\mathcal{L}} \subset L_2(\mathcal{M}, d\mu)$, and an isomorphism Ξ between $\check{\mathcal{L}}$ and $\mathcal{H}$,

(c) for each relevant observable $\hat{A}$ a function $f_A : \mathcal{M} \to \mathcal{C}$ such that the Toeplitz quantization of f_A yields the operator A: $\Xi\, T_{f_A} \Xi^{-1} = \hat{A}$,

(d) a function H_{cl} such that the Toeplitz quantization of $G_{H_{cl}}$ yields the operator $\hat{H}$.

We remark that dequantization is not a classical limit procedure involving $\hbar \to \infty$ but a correpondence, i.e. a map that assigns a classical system to a given quantum system.

We remark that the dequantization of the Hamiltonian, i.e. of the generator of the quantum dynamics, is different than the dequantization of the observables. A procedure of dequantization along the above requirements can be formulated using coherent states as follows.

6.0.1. (a) Construction of a phase space manifold from a set of coherent states

The first step is the construction of a set of coherent states. For any given quantum system, the choice of coherent states is not unique. The approach of Gilmore and of Perelomov, based on a group theoretical construction, yields a phase space manifold $\mathcal{M}$, a measure $d\mu$ and a set of unnormalized coherent states satisfying the completeness relation $\int d\mu(\underline{z}_0)\, |\zeta_{\underline{z}_0}\rangle\langle\zeta_{\underline{z}_0}| = \mathbb{1}$. This step is described in detail in [20, 23, 41].

6.0.2. (b) Construction of the polarization subspace and of the isomorphism Ξ from a given set of unnormalized coherent states

If a complete set of unnormalized coherent states $\{|\zeta_{\underline{z}_0}^{\mathcal{H}}\rangle \in \mathcal{H}\}_{\underline{z}_0 \in \mathcal{M}}$ is given, one can construct a polarization subspace $\check{\mathcal{L}} \subset \mathcal{L}_K$ and an isomorphism $\Xi : \check{\mathcal{L}} \to \mathcal{H}$ such that the states $|\zeta_{\underline{z}_0}\rangle$ defined by Eq. (5.53) coincide with the states $|\zeta_{\underline{z}_0}^{\mathcal{H}}\rangle$:

$$\Xi|\zeta_{\underline{z}_0}\rangle = |\zeta_{\underline{z}_0}^{\mathcal{H}}\rangle.$$

This can be shown as follows. We introduce a map $\Xi_{Hus} : \mathcal{H} \to \mathcal{L}_K$ by

$$\Xi_{Hus} : \psi \mapsto \xi \quad \text{defined by} \quad \xi(\underline{z}_0) := \langle\zeta_{\underline{z}_0}^{\mathcal{H}}|\psi\rangle.$$

We will call Ξ_{Hus} the *Husimi map* since $|\xi(\underline{z}_0)|^2$ is the Husimi function corresponding to the state ψ. The Husimi map defines an isomorphism between $\mathcal{H}$ and a subspace $\check{\mathcal{L}}$ of $\mathcal{L}_K$, that satisfies the following properties:

(i) Ξ_{Hus} is a continuous linear map that preserves the scalar products, i.e.

$$\langle\, \Xi_{Hus}(\psi_1) \mid \Xi_{Hus}(\psi_2)\,\rangle_{\mathcal{L}_K} = \langle\psi_1|\psi_2\rangle_{\mathcal{H}},$$

since

$$\langle\, \Xi_{Hus}(\psi_1) \mid \Xi_{Hus}(\psi_2)\,\rangle_{\mathcal{L}_K} = \int_{\mathcal{M}} d\mu(\underline{z}_0)\, (\Xi_{Hus}(\psi_1))^* \Xi_{Hus}(\psi_2)$$

$$= \int_{\mathcal{M}} d\mu(\underline{z}_0)\, \langle\psi_1|\zeta_{\underline{z}_0}^{\mathcal{H}}\rangle_{\mathcal{H}} \langle\zeta_{\underline{z}_0}^{\mathcal{H}}|\psi_2\rangle_{\mathcal{H}}$$

$$= \langle\psi_1| \left(\int_{\mathcal{M}} d\mu(\underline{z}_0)\, |\zeta_{\underline{z}_0}^{\mathcal{H}}\rangle\langle\zeta_{\underline{z}_0}^{\mathcal{H}}| \right) |\psi_2\rangle$$

$$= \langle\psi_1|\psi_2\rangle_{\mathcal{H}}.$$

This implies that the image of $\psi \in \mathcal{H}$ is indeed in the space $\mathcal{L}_K$ of square-integrable functions.

(ii) The image $\Xi_{Hus}(\mathcal{H}) =: \check{\mathcal{L}}$ is a subspace of $\mathcal{L}_K$.

(iii) We chose an arbitrary orthonormal basis $\{|n\rangle\}_{n \in I \subset \mathbb{Z}}$ of $\mathcal{H}$. Its image by the Husimi map defines

$$\eta_n := \Xi_{Hus}|n\rangle.$$

The set of functions $\{\eta_n\}_{n \in I \subset \mathbb{Z}}$ is an orthonormal set that spans the subspace $\check{\mathcal{L}} \subset \mathcal{L}_K$. If we define the isomorphism $\Xi : \check{\mathcal{L}} \to \mathcal{H}$ by $\eta_n \mapsto |n\rangle$, we can identify it as the inverse of the Husimi map : $\Xi = \Xi_{Hus}^{-1}$, and

$$\Xi|\zeta_{\underline{z}_0}\rangle = \Xi \sum_n \eta_n^*(\underline{z}_0)\, |\eta_n\rangle$$

$$= \sum_n \eta_n^*(\underline{z}_0)\, |n\rangle = |\zeta_{\underline{z}_0}^{\mathcal{H}}\rangle.$$

We consider as an example the case of the standard Glauber coherent states defined on the Fock space $\mathcal{H}$ by

$$|C_{\underline{z}}^{\mathcal{H}}\rangle := e^{z\hat{a}^\dagger - z^*\hat{a}}|n = 0\rangle.$$

With respect to the basis $\{|n\rangle\}$ they are expressed as

$$|C_{\underline{z}}^{\mathcal{H}}\rangle = e^{-zz^*/2} \sum_{n=0}^{\infty} \frac{z^n}{\sqrt{n!}}|n\rangle. \tag{6.1}$$

Since they satisfy the completeness relation

$$\mathbb{1} = \frac{1}{\pi} \int d^2z |C_{\underline{z}}^{\mathcal{H}}\rangle\langle C_{\underline{z}}^{\mathcal{H}}| = \frac{1}{2\pi\hbar} \int d\mu(\underline{z})|C_{\underline{z}}^{\mathcal{H}}\rangle\langle C_{\underline{z}}^{\mathcal{H}}|,$$

the unnormalized coherent states are

$$|\zeta_{\underline{z}}^{\mathcal{H}}\rangle = \frac{1}{\sqrt{2\pi\hbar}}|C_{\underline{z}}^{\mathcal{H}}\rangle.$$

The image of the corresponding Husimi map $\Xi_{Hus} : \mathcal{H} \to \mathcal{L}_K$ is spanned by the following vectors:

$$\Xi_{Hus}|n\rangle \equiv \langle\zeta_{\underline{z}}^{\mathcal{H}}|n\rangle = \frac{1}{\sqrt{2\pi\hbar n!}}z^n e^{-zz^*/2}. \tag{6.2}$$

This subspace is different from the polarization subspace $\check{\mathcal{L}}$ we chose in (5.12), since z^n generates holomorphic instead of anti-holomorphic functions. In order to obtain the subspace $\check{\mathcal{L}}$ we have to choose a slightly different set of coherent states, exchanging z and z^*:

$$|C'^{\mathcal{H}}_{\underline{z}}\rangle := e^{z^*\hat{a}^\dagger - z\hat{a}}|n=0\rangle, \qquad |\zeta'^{\mathcal{H}}_{\underline{z}}\rangle = \frac{1}{\sqrt{2\pi\hbar}}|C'^{\mathcal{H}}_{\underline{z}}\rangle. \tag{6.3}$$

which leads to the subspace generated by

$$\Xi'_{Hus}|n\rangle \equiv \langle\zeta'^{\mathcal{H}}_{\underline{z}}|n\rangle = \frac{1}{\sqrt{2\pi\hbar n!}}z^{*n} e^{-zz^*/2}, \tag{6.4}$$

which is equal to the polarization subspace $\check{\mathcal{L}} \in L_K$ of the Berezin-Toeplitz quantization that we defined in (5.12).

6.0.3. (c) Dequantization of the observables — covariant and contravariant symbols

For a given operator $\hat{A}$ one can define two types of *symbols*, which are functions or more generally distributions on the phase space $\mathcal{M}$:

(i) The *contravariant symbol* $f_{\hat{A}}$ — also called *upper bound symbol* or *P-symbol* — is defined as a function (or more generally distribution) such that $\hat{A} = \hat{T}_{f_{\hat{A}}}$, i.e.

$$\hat{A} = \hat{T}_{f_A} \equiv \Xi \, P_{\check{\mathcal{L}}} f_{\hat{A}} P_{\check{\mathcal{L}}} \, \Xi^{-1} \equiv \int_{\mathcal{M}} d\mu(\underline{z}_0) f_{\hat{A}}(\underline{z}_0) \, |\zeta_{\underline{z}_0}^{\mathcal{H}}\rangle\langle\zeta_{\underline{z}_0}^{\mathcal{H}}|. \tag{6.5}$$

(ii) The *covariant symbol* $S_{\hat{A}}$ — also called *lower bound symbol* or *Q-symbol* — is defined as

$$S_{\hat{A}}(\underline{z}_0) := \langle\zeta_{\underline{z}_0}^{\mathcal{H}}|\hat{A}|\zeta_{\underline{z}_0}^{\mathcal{H}}\rangle. \tag{6.6}$$

We remark that while for some operators $\hat{A}$ the symbols can be expected to be smooth functions on $\mathcal{M}$ for other operators the symbols may not be well-defined or they may be a more singular object like e.g. a distribution.

The validity of the following formal relations between an operator $\hat{A}$ and its covariant and contravariant symbols must be analyzed for each particular type of operator.

(1) The covariant Q-symbol $S_{\hat{A}}(z_0)$ can be calculated directly, provided that the coherent states are in the domain of definition of the operator $\hat{A}$, i.e. provided that the scalar product is well defined. For the standard Glauber coherent states, it can be expressed also as [17, 23]

$$S_{\hat{A}}(z_0) = \frac{1}{\pi} \int_{\mathbb{C}} d^2z \; e^{z_0 z^* - z_0^* z} \; \text{Tr}(\hat{A}e^{-z^* \hat{a}} e^{z \hat{a}^\dagger}). \tag{6.7}$$

(2) The contravariant P-symbol $f_{\hat{A}}(z_0)$ can be written as the following formal expression:

$$f_{\hat{A}}(z_0) = \frac{1}{\pi} \int_{\mathbb{C}} d^2z \; e^{z_0 z^* - z_0^* z} \; \text{Tr}(\hat{A}e^{z \hat{a}^\dagger} e^{-z^* \hat{a}}). \tag{6.8}$$

The covariant symbol $S_{\hat{A}}$ is generally easier to calculate and more regular that the covariant one $f_{\hat{A}}$. $f_{\hat{A}}$ can be expressed in terms of $S_{\hat{A}}$ through their Fourier transforms: defining the Fourier transforms

$$\tilde{f}_{\hat{A}}(w) := \frac{1}{\pi} \int_{\mathbb{C}} d^2z \; f_{\hat{A}}(z) e^{wz^* - w^* z}, \tag{6.9}$$

$$\tilde{S}_{\hat{A}}(w) := \frac{1}{\pi} \int_{\mathbb{C}} d^2z \; S_{\hat{A}}(z) e^{wz^* - w^* z} \tag{6.10}$$

and their inverses

$$f_{\hat{A}}(z) := \frac{1}{\pi} \int_{\mathbb{C}} d^2z \; \tilde{f}_{\hat{A}}(w) e^{-wz^* + w^* z}, \tag{6.11}$$

$$S_{\hat{A}}(z) := \frac{1}{\pi} \int_{\mathbb{C}} d^2z \; \tilde{S}_{\hat{A}}(w) e^{-wz^* + w^* z} \tag{6.12}$$

one can establish [17, 23] the relation

$$\tilde{f}_{\hat{A}}(w) = e^{w^* w} \tilde{S}_{\hat{A}}(w),$$

which, applying the inverse Fourier transform, can be written as

$$f_{\hat{A}}(z) = e^{-\frac{\partial^2}{\partial z \partial z^*}} S_{\hat{A}}(z) = e^{-\frac{\partial^2}{\partial z \partial z^*}} \langle \zeta_z^{\mathcal{H}} | \hat{A} | \zeta_z^{\mathcal{H}} \rangle. \tag{6.13}$$

Remark. This relation is also true for the conjugate Glauber coherent states (6.3), since it is invariant upon the exchange of z an z^*.

6.0.4. (d) Dequantization of the Hamiltonian generator of the dynamics

Given an operator $\hat{H}$ in $\mathcal{H}$, we want to determine a function $H(p,q)$, such that

$$\hbar \widehat{G}_H = \hat{H}.$$

Using Tuynman's relation, this is equivalent to

$$\widehat{T}_{\tau(H)} = \hat{H}$$

The function $h := \tau(H)$ is, by definition, the contravariant symbol of $\hat{H}$, which according to Eqs. (6.6), (6.13) can be expressed as

$$h = e^{-\frac{\partial^2}{\partial z \partial z^*}} \langle \zeta_z^{\mathcal{H}} | \hat{H} | \zeta_z^{\mathcal{H}} \rangle. \tag{6.14}$$

In order to obtain the function H we have to invert Tuynman's relation (5.37):

$$H - \frac{\partial^2 H}{\partial z \partial z^*} = h,$$

which we can write formally as

$$H = \left(1 - \frac{\partial^2}{\partial z \partial z^*} \right)^{-1} h.$$

Inserting (6.14) we obtain

$$H = \left(1 - \frac{\partial^2}{\partial z \partial z^*} \right)^{-1} e^{-\frac{\partial^2}{\partial z \partial z^*}} \langle \zeta_z | \hat{H} | \zeta_z \rangle. \tag{6.15}$$

We remark that $-\frac{\partial^2 f}{\partial z \partial z^*} = -\frac{\hbar}{2} \left(\frac{1}{\beta_0} \frac{\partial^2 f}{\partial q^2} + \beta_0 \frac{\partial^2 f}{\partial p^2} \right)$ is a positive operator ($\sim -\Delta$ in adapted coordinates). Thus $(1 - \frac{\partial^2}{\partial z \partial z^*})^{-1}$ is well defined and bounded in a suitably defined function space [42]. However, $e^{-\frac{\partial^2}{\partial z \partial z^*}}$ is an unbounded operator, which is the origin of the regularity difficulties of the contravariant symbol. It will be regular if the contravariant symbol is in the domain of the Laplacian.

7. Conclusions

In summary, the formalism that we have described allows one to construct models describing the interaction between classical and quantum systems in a well-defined Hilbert space framework. The geometric quantization of a classical system consists of selecting a subspace of the classical Hilbert space of functions on phase space. The quantization of the observables is defined by projecting the classical observables on this subspace. The quantization of the dynamics involves first the addition of a dynamical and a geometrical phase to the classical dynamics and then projecting the generator of the dynamics on the subspace. The dequantization of a quantum model consists of the inverse procedure: given a Hamiltonian, an algebra of observables represented in a Hilbert space, and a set of coherent states, one can construct an associated phase space manifold and the classical Hilbert space of square-integrable functions, with a suitable subspace that gives back the original quantum model when the geometric quantization is performed.

In the definition of the quantum models by Berezin-Toeplitz-geometrical quantization, Planck's constant $\hbar$ appears in two places, that can be considered conceptually independent: The first one is in the phase factor (3.3) of the pre-quantum Koopman-Schrödinger wave function. The second one is in the selection of the polarization subspace, which depends crucially on the value of the constant $\hbar$. Although in principle the two constants could be taken with two different independent values (to be determined by comparison with experiments), they are taken to be equal to a single constant $\hbar$.

Aknowledgments

We acknowlege support of the Marie Curie ITN Network FASTQUAST. We thank M. Lachièze-Rey, F. Faure and S. de Bièvre for very helpful discussions.

References

1. B. O. Koopman; Hamiltonian systems and transformations in Hilbert spaces; Proc. Natl. Acad. Sci. USA 17 (1931) 315-318.
2. J. von Neumann; Zur Operatorenmethode in der klassischen Mechanik; Ann. Math. 33 (1932) 587-642; ibid. 33 (1932) 789.
3. E. Deotto, E. Gozzi and D. Mauro; Hilbert space structure in classical mechanics. I, II; J. Math. Phys. 44 (2003) 5902-5936 and 5937-5957.
4. N. M. J. Woodhouse; Geometric Quantization; Clarendon Press, Oxford, 1992.
5. J. M. Souriau; Structure des systèmes dynamiques; Dunod, Paris 1970.
6. B. Kostant; Quantization and unitary representations; Springer Lecture Notes in Math. 170 (1970) 85-208.
7. F. Faure; Prequantum chaos: resonances of the prequantum cat map; Journal of Modern Dynamics 1 (2007)255.

8. F. Faure; Exposé sur la quantification géometrique; unpublished lecture notes, Institut Fourier, Grenoble (2000).

9. G. A. Gallup, H. Batelaan and T. J. Gay; Quantum-mechanical analysis of a longitudinal Stern-Gerlach effect; Phys. Rev. Lett. 86 (2001) 4508.

10. G. Potel, F. Barranco, S. Cruz-Barrios and J. Gomez-Camacho; Quantum mechanical description of Stern-Gerlach experiments; Phys. Rev. A 71 (2005) 052106.

11. S. Cruz-Barrios and J. Gomez-Camacho; Semiclassical description of Stern-Gerlach experiments; Phys. Rev. A 63 (2000) 012101.

12. I. Bialynicki-Birula and Z. Bialynicka-Birula; Quantum electrodynamics of intense photon beams: New approximation method; Phys. Rev. A 14 (1976) 1101.

13. I. Bialynicki-Birula and C. L. Van; Energy levels of dressed atoms and resonance phenomena; Acta Phys. Pol. A 57 (1980) 599.

14. S. Guérin, F. Monti, J.-M. Dupont and H. R. Jauslin; On the relation between cavity dressed states, Floquet states, RWA approximation and semiclassical models; J. Phys. A 30 (1997) 7193.

15. J. R. Klauder; The Current State of Coherent States; arXiv:quant-ph/0110108 (2001).

16. J. R. Klauder; Continuous representation theory II. Generalized relation between quantum and classical dynamics; J. Math. Phys. 4 (1963) 1058-1073.

17. J. R. Klauder and B. S. Skagerstam; Coherent States: Applications in Physics and Mathematical Physics; World Scientific, Singapore, 1985.

18. E. Schrödinger; Der stetige Übergang von der Mikro-zur Makromechanik; Naturwiss. 14 (1926) 664.

19. R. J. Glauber; Coherent and incoherent states of the radiation field; Phys. Rev. 131 (1963) 2766.

20. A. M. Perelomov; Generalized Coherent States and Their Applications; Springer Verlag, Berlin, 1986.

21. A. M. Perelomov; Coherent states for arbitrary Lie groups; Commun. Math. Phys. 26 (1972) 222-236.

22. R. Gilmore; On the properties of coherent states; Rev. Mexicana de Fisica 23 (1974) 143-187.

23. W.-M. Zhang, D. H. Feng and R. Gilmore; Coherent states: Theory and some applications; Rev. Mod. Phys. 62 (1990) 867.

24. F. T. Arecchi, E. Courtens, R. Gilmore and H.Thomas; Atomic coherent states in quantum optics; Phys. Rev. A 6 (1972) 2211.

25. A. O. Barut and L. Girardello; New coherent states associated with non-compact groups; Commun. Math. Phys. 21 (1972) 41-55.

26. M. M. Nieto and L. M. Simmons; Coherent states for general potentials; Phys. Rev. Lett. 41 (1978) 207-210.

27. S. T. Ali, J.-P. Antoine and J.-P. Gazeau; Coherent States, Wavelets and their Generalizations; Springer-Verlag, New York (2000).

28. G. S. Agarwal and E. Wolf; Phys. Rev. D 2 (1970) 2161; 2187; 2206.

29. M. Hillery, R. F. O'Connell, M. O. Scully and E. P. Wigner; Distribution functions in physics: Fundamentals; Phys. Rep. 106 (1984) 121-167.

30. V. Bargmann; On a Hilbert space of analytic functions and an associated integral transform, Part I; Commun. Pure Appl. Math. 14 (1961) 187-214.

31. V. Bargmann; Remarks on a Hilbert space of analytic functions; Proc. Nat. Academy Sci. USA 48 (1962) 199-204.

32. B. Hall; Holomorphic Methods in Analysis and Mathematical Physics; arXiv:quant-ph/9912054.

33. M. Lachièze Rey, J.-P. Gazeau, T. Garidi, E. Huguet and J. Renaud; Quantization of the sphere with coherent states; Int. J. Theor. Phys. 42 (2003) 1301-1310.

34. J.-P. Gazeau, T. Garidi, E. Huguet, M. Lachièze Rey and J. Renaud; Examples of Berezin-Toeplitz Quantization: Finite sets and Unit Interval; Proceedings of the Workshop in honor of R. Sharp, Montreal 2002, P. Winternitz ed., CRM-AMS (2004) p. 67-76, (http://arXiv.org/abs/quant-ph/0303090).

35. J.-P. Gazeau, E. Huguet and M. Lachièze-Rey; Fuzzy spheres from inequivalent coherent states quantizations; J. Phys. A 40 (2007) 10225–10249.

36. S. Twareque Ali, J.-P. Antoine, J.-P. Gazeau and U. A. Mueller; Coherent states and their generalizations: a mathematical overview; Rev. Math. Phys. 7 (1995) 1013–1104.

37. S. Twareque Ali and J.-P. Antoine; Quantum frames, quantization and dequantization; in Quantization and infinite-dimensional systems (Bialowieza, 1993), pp. 133-145, Plenum, New York, 1994.

38. S. Twareque Ali and M. Engliš; Quantization methods: a guide for physicists and analysts; Reviews in Mathematical Physics 17 (2005) 391–490.

39. M. Reed and B. Simon; Methods of Modern Mathematical Physics I; Academic Press, San Diego, 1975.

40. J. M. Gracia-Bondia; Generalized Moyal quantization on homogeneous symplectic spaces, in Deformation theory and quantum groups with applications to mathematical physics (Amherst, MA, 1990); pp. 93-114, Comtemp. Math. vol 134, AMS, Providence, 1992.

41. W.-M. Zhang and D. H. Feng; Quantum nonintegrability in finite systems; Phys. Rep. 252 (1995) 1–100.

42. G. M. Tuynman; Quantization: Towards a comparison between methods; J. Math. Phys. 28 (1987) 2829-2840.

43. C. Duval, J. Elhadad and G. M. Tuynman; Hyperfine interaction in a classical hydrogen atom and geometric quantization; J. Geom. Phys. 3 (1986) 401.

44. M. Bordemann, J. Hoppe, P. Schaller and M. Schlichenmaier; $gl(\infty)$ and geometric quantization; Commun. Math. Phys. 138(1991), 209-244.

45. M. Bordemann, E. Meinrenken and M. Schlichenmaier; Toeplitz Quantization of Kähler Manifolds and $gl(N)$, $N \rightarrow \infty$; Commun. Math. Phys. 165 (1994) 281-296.

QUANTUM MEMORIES AS OPEN SYSTEMS

Robert Alicki

Institute of Theoretical Physics and Astrophysics
University of Gdańsk, PL 80-952 Gdańsk, Poland
E-mail: fizra@univ.gda.pl

Promising candidates for quantum memory are N-spin systems with specially designed Hamiltonians weakly coupled to heat baths. Their dynamics can be described by quantum dynamical semigroups of the Davies type. The rigorous results concerning 2D and 4D Kitaev models are briefly reviewed and their physical meaning is discussed.

0. Introduction

Any attempt of large scale implementations of the idea of quantum computing demands fault-tolerance. The existing theory of fault-tolerant quantum computing (FTQC) with the famous threshold theorems [1, 2] is based on the phenomenological assumptions which disagree with the fundamental features of the Hamiltonian approach to quantum computation [3, 4]. On the other hand the rigorous analysis of Hamiltonian models of computers executing generic quantum algorithms is still out of range. Therefore, it is reasonable to restrict the discussion to the problem of the existence of quantum memory (QM), i.e. of a device which can preserve an unknown M-qubit state for a sufficiently long time. One should also have an efficient methods of preparation and readout of such a state. In any case, the most important feature of QM is its scalability. Firstly, one should be able to compose a M-qubit memory from M units (1-qubit QMs) each of them consisting of N physical qubits. Secondly, under achievable external conditions (low enough temperature, high vacuum, screening , ..., etc.) the life-time of encoded qubit observables should increase exponentially with N. The last condition is necessary to allow arbitrary long effective quantum computations. In the following by QM we mean such a scalable, exponentially stable 1-qubit QM.

It is obvious that the idea of QM presents a challenge to the Bohr Correspondence Principle (BCP). In its general formulation BCP demands that:

Classical physics and quantum physics give the same answer when the system become large,
or in other words:
For large systems the experimental data are consistent with classical probabilistic models.

Indeed, assuming that BCP is universally valid, large N QM device should be described by a classical model and therefore cannot carry a genuine quantum information.
Therefore the natural question arises:
Are there fundamental obstacles to build a Quantum Memory?

Although some more or less heuristic "no-go theorems" for QM have been presented [5, 6], the rigorous analysis of the proposed models for QM is inevitable to clarify this issue. All models considered in the following are interacting systems of N physical qubits called spins weakly coupled to a heat bath at the temperature $T > 0$. Although, there exist different sources of noise, only the thermal fluctuations cannot be completely eliminated by a proper "engineering". Under certain conditions on the system Hamiltonian and its coupling to a heat bath the dynamics of the N-spin systems is governed by the quantum Markovian master equation (QMME) in the Davies form [7] (a particular case of the Lindblad-Gorini-Kossakowski-Sudarshan QMME [8, 9]). For such models the question of existence of stable encoded qubit reduces to the the spectral analysis of the corresponding Davies generator in the Heisenberg picture. This leads to interesting mathematical problems which were treated in the papers [10–12] for the particular models: 1D Ising, and 2, 3, 4D Kitaev models. The Kitaev models were considered as relatively simple candidates for QM [13, 14]. We shall briefly discuss the obtained rigorous results. While the 2D and 3D Kitaev models do not support stable encoded qubits the 4D case possesses exponentially stable qubit observables. However, the effective manipulations with such an encoded qubit are questionable.

1. Encoded Qubit

By a *physical qubit* we mean a "natural" system described by 2-dim Hilbert space and the algebra $\mathcal{M}_2$ of 2×2 matrices spanned by I, σ^x, σ^y, σ^z.

Sometimes this description is exact like for spin-1/2 or photon's polarization, in other cases only approximated like for a "2-level atom" or mesoscopic bistable systems. An *encoded qubit* can be identified with a subalgebra $\mathcal{Q}$ of the algebra of observables $\mathcal{A}$ of the total system spanned by the self-adjoint elements I, X, Y, Z satisfying $X^2 = Y^2 = Z^2 = I$, $XY = iZ$, and cyclic permutations. Equivalently $\mathcal{Q}$ can be seen as a subalgebra generated by $X, Z \in \mathcal{A}$ such that

$$X = X^\dagger, \quad Z = Z^\dagger, \quad X^2 = Z^2 = I, \quad XZ + ZX = 0. \qquad (1.1)$$

In the following the total system is a N-spin system with the $2^N \times 2^N$ matrix algebra $\mathcal{A} = \mathcal{M}_{2^N}$. As an illustration, consider the following examples:

(1) Trivial encoding into spin

$$X = \sigma_1^x \otimes I_{[2,N]}, \qquad Z = \sigma_1^z \otimes I_{[2,N]}. \qquad (1.2)$$

(2) Encoding for 1D Ising model with periodic boundary conditions and the Hamiltonian

$$H_N^{1D} = -J \sum_{j=1}^{N} \sigma_j^z \sigma_{j+1}^z = -J \sum_{j=1}^{N} b_j, \qquad (\sigma_{N+1}^z \equiv \sigma_1^z) \qquad (1.3)$$

written in terms of *bonds* $b_j = \sigma_j^z \sigma_{j+1}^z$. We define two examples of encoded qubit generators commuting with H_N^{1D}:

$$X = \sigma_1^x \otimes \sigma_2^x \otimes \cdots \sigma_N^x, \qquad Z = \sigma_1^z \otimes \sigma_2^x \otimes \cdots \sigma_N^x, \qquad (1.4)$$

$$X' = XF_x, \qquad Z' = ZF_z, \qquad F_{x,z}^\dagger = F_{x,z}, \\ F_{x,z}^2 = I, \qquad F_{x,z}\text{-function of bonds}. \qquad (1.5)$$

One can show that the example (1.5) is the most general encoded qubit commuting with all bonds. It means that such qubit observables are constant of motion for all Hamiltonians which are functions of bonds. A similar construction will be used for Kitaev models also.

2. A Generic Model of Classical Memory

In order to understand the mechanism of information protection against thermal noise we begin with the brief discussion of the classical case. Typically, to encode classical information metastable local minima of free energy separated by free energy barriers are used. As the height F of such a barrier is proportional to the size of the system given by the number of its microscopic constituents N the probability of transition between local minima is dominated by the Boltzmann factor $\exp\{-F/kT\}$ what leads to exponentially long (in N) life-times.

As paradigmatic examples one can consider classical Ising models used to encode a single bit — magnetization's sign. To see what are the necessary conditions to achieve stability we compare a mean-field Ising model with the Hamiltonian

$$H_N^{mf} = -\frac{J}{2N} \sum_{i,j=1}^{N} \sigma_i^z \sigma_j^z \qquad (2.1)$$

with the 1D one given by (1.3). Comparing the energy difference between two spin configuration:

$$+ + + + + + + + + + + + \quad \text{and} \quad + + + \underbrace{- - - - - -}_{k\text{-times}} + + + + + + ,$$

one obtains

$$\Delta E^{mf} = Jk - \frac{Jk^2}{2N} , \qquad \Delta E^{1D} = 2J. \qquad (2.2)$$

For the mean-field model the energy difference grows with the number of flipped spins what provides a mechanism of bit's protection against noise and leads to the phase transition phenomenon below the critical temperature. On the other hand for 1D Ising model only finite portion of energy is needed to reverse all spins and hence no phase transition is present.

A more detailed analysis of the stability for classical lattice models involves stochastic dynamics which simulates the evolution of the system coupled to a heat bath. The standard example is the Glauber dynamics for classical Ising models.

Denote by $s = \{\sigma_k = \pm 1, k = 1, 2, \ldots, N\}$ the configuration of N spins, and by s^j the configuration s with one "j"-spin flipped, $\sigma_j \to -\sigma_j$. The time-dependent probability distribution on the set of configurations denoted by $P_s(t)$ satisfies the following Markovian Master Equation

$$\frac{d}{dt} P_s = \gamma \sum_{j=1}^{N} \left(P_{s^j} - e^{-\frac{E_j(s)}{kT}} P_s \right) \qquad (2.3)$$

where $E_j(s) = H_N(s^j) - H_N(s)$ and γ is the relaxation rate. The Glauber dynamics possesses several important properties which are consistent with the phenomenology of thermal relaxation:

(1) Equilibrium Gibbs state

$$P_s^{eq} = Z^{-1} e^{-\frac{H_N(s)}{kT}} \tag{2.4}$$

is a stationary solution of (2.3),

(2) For any initial state $P_s(0)$ the solution of (2.3) converges to P_s^{eq}

$$\lim_{t\to\infty} P_s(t) = P_s^{eq}, \tag{2.5}$$

(3) The detailed balance condition holds, i.e. at equilibrium the probability flow from $'s$ to s^j is exactly compensated by the reverse process.

There exist numerous results for Glauber dynamics, both rigorous and numerical supporting the picture presented above of metastable exponentially long living states below the critical temperature for models exhibiting phase transitions. In order to check whether a similar mechanism of metastability can be used to protect a quantum state one has to propose suitable N-spin quantum models with quantum evolutions corresponding to Glauber dynamics.

3. Davies Generators

Consider a quantum N-spin system with the Hamiltonian H_N weakly coupled to a heat bath at the temperature $T > 0$. The structure of the coupling is fundamental for the discussion of stability. A generic coupling can be seen as a combination of two extreme cases: collective bath coupling and private baths one. The interaction Hamiltonian of the collective type has form

$$H_{int} = \sum_{\mu=x,y,z} \left(\sum_{j=1}^{N} \sigma_j^\mu \right) \otimes F^\mu \tag{3.1}$$

and is invariant with respect to spin permutations. This invariance implies the existence of decoherence-free subsystems corresponding to some subalgebras of observables which are not affected by the environment [15, 16]. Physically, such a symmetry with respect to permutations can be treated only as a rough approximation (compare the phenomenon of superradiance). Real systems possess certain ergodic properties which forbid the existence of decoherence-free subsystems. Such ergodicity is realized by the private heat baths coupling of the form

$$H_{int} = \sum_{\mu=x,y,z} \sum_{j=1}^{N} \sigma_j^\mu \otimes F_j^\mu, \tag{3.2}$$

which will be used here as a simplifying assumption. Moreover, we assume that all heat baths are identical as well as all baths' observables $F_j^\alpha \equiv F$. Therefore, also $H_B = \sum H_B^{\mu,j}$ and a bath's equilibrium state $\omega_B = \otimes \omega_B^{\mu,j}$ with identical copies of $H_B^{\mu,j}$ and $\omega_B^{\mu,j}$.

The starting point for the analysis is the reduced dynamics for the density matrix of the spin system given in terms of a partial trace over the bath

$$\rho(t) = \mathrm{Tr}_B\big(U(t)\rho(0) \otimes \omega_B U^\dagger(t)\big) \tag{3.3}$$

where $U(t)$ is a unitary dynamics of the total system governed by the Hamiltonian

$$H_{NB} = H_N + H_B + \lambda \sum_\alpha S_\alpha \otimes F_\alpha. \tag{3.4}$$

Strictly speaking the *physical* Hamiltonian H_N in (3.4) should be replaced by a *bare* Hamiltonian which contains some *counterterms* used to cancel the Hamiltonian corrections caused by the interaction with the bath. We do not go into details of this renormalization procedure and in the final formulas concerning QMMEs only physical Hamiltonian appears. In the interaction Hamiltonian the explicit small coupling constant λ is present, and S_α is a shorthand notation for $\sigma_j^{x,y,z}$ from Eq. (3.2). The standard assumption

$$\mathrm{Tr}(\omega_B F_\alpha) = 0 \tag{3.5}$$

is also added.

Denote by $\{\omega\}$ the set of Bohr frequencies of the Hamiltonian H_N (i.e. all differences of its eigenvalues), and let $S_\alpha(\omega)$ be the discrete Fourier components of S_α in the interaction picture, i.e.,

$$S_\alpha(t) = \exp(iH_N t)S_\alpha \exp(-iH_N t) = \sum_{\{\omega\}} S_\alpha(\omega) \exp(i\omega t). \tag{3.6}$$

A well-known sequence of approximations involving van Hove *weak coupling limit*, discussed for example in [7, 17] leads to the following QMME

$$\frac{d\rho}{dt} = -i[H_N, \rho] + \mathcal{L}\rho, \tag{3.7}$$

$$\mathcal{L}\rho \equiv \frac{1}{2}\lambda^2 \sum_\alpha \sum_{\{\omega\}} R(\omega)\Big([S_\alpha(\omega), \rho S_\alpha(\omega)^\dagger] + [S_\alpha(\omega)\rho, S_\alpha(\omega)^\dagger]\Big). \tag{3.8}$$

Here

$$R(\omega) = \int_{-\infty}^{\infty} e^{i\omega t} \text{Tr}(\omega_B F(t) F) \, dt \qquad (3.9)$$

where $F(t) = e^{iH_B t} F e^{-iH_B t}$. The spectral density $R(\omega)$ at equilibrium state satisfies the Kubo-Martin-Schwinger condition

$$R(-\omega) = e^{-\omega/kT} R(\omega). \qquad (3.10)$$

For the further discussion the Heisenberg picture version of the evolution (3.7), (3.8) is more convenient

$$\frac{dA}{dt} = i\mathcal{H}A + \mathcal{L}^* A, \quad \mathcal{H}A \equiv [H_N, A] \qquad (3.11)$$

where

$$\mathcal{L}^* A \equiv \frac{1}{2}\lambda^2 \sum_{\alpha} \sum_{\{\omega\}} R(\omega)\left(S_\alpha(\omega)^\dagger [A, S_\alpha(\omega)] + [S_\alpha(\omega)^\dagger, A] S_\alpha(\omega)\right). \qquad (3.12)$$

The sum $\mathcal{G} = i\mathcal{H} + \mathcal{L}^*$ generates a semi-group of completely positive, identity preserving transformations on the algebra of observables with certain additional properties due to the weak coupling limit construction and the KMS condition:

(D1) The canonical Gibbs state is stationary with respect to (3.7,3.8)

$$\text{Tr}\left(\rho^{eq} e^{t\mathcal{G}}(X)\right) = \text{Tr}\left(\rho^{eq} X\right) \qquad (3.13)$$

where

$$\rho^{eq} = \frac{e^{-H_N/kT}}{\text{Tr}(e^{-H_N/kT})}. \qquad (3.14)$$

(D2) The semi-group is relaxing, i.e. any initial state ρ evolves to ρ^{eq}

$$\lim_{t \to \infty} \text{Tr}\left(\rho e^{t\mathcal{G}}(X)\right) = \text{Tr}(\rho^{eq} X), \qquad (3.15)$$

(D3) $\mathcal{L}^*$ satisfies the *quantum detailed balance condition*

$$\mathcal{H}\mathcal{L}^* = \mathcal{L}^* \mathcal{H} \qquad (3.16)$$

and

$$\text{Tr}\left(\rho^{eq} Y^\dagger \mathcal{L}^*(X)\right) = \text{Tr}\left(\rho^{eq} \left(\mathcal{L}^*(Y)\right)^\dagger X\right). \qquad (3.17)$$

Equation (3.17) expresses the self-adjointness of $\mathcal{L}^*$ with respect to the Liouville scalar product

$$\langle X, Y \rangle_{eq} := \text{Tr}\left(\rho^{eq} X^\dagger Y\right). \qquad (3.18)$$

(D4) The dissipative part $\mathcal{L}^*$ of the generator is negative definite.

(D5) Spectral decomposition yields the orthonormal modes X_μ, $\nu = 0, 1, \ldots, 2^{2N} - 1$ decomposition

$$A(t) = \sum_\nu e^{(i\omega_\nu - \lambda_\nu)t} \langle X_\nu, A(0) \rangle_{eq} X_\nu, \quad \lambda_\nu > 0 \qquad (3.19)$$

with $X_0 \equiv I$, $i\omega_0 - \lambda_0 = 0$ the only eigenvalue equal to zero.

(D6) The diagonal elements of the density matrix computed in the Hamiltonian basis evolve independently of the off-diagonal ones and their evolution is governed by a kind of Glauber dynamics.

Due to (D2) any initial state of the system will eventually relax to equilibrium. However, it does not exclude the existence of metastable states with life-times exponentially growing with N. The existence of the corresponding metastable observables which could encode qubits depends on the properties of the lowest non-zero eigenvalue of $-\mathcal{L}^*$ called *spectral gap*.

4. Kitaev Models

The Kitaev models in $D = 2, 3, 4$ dimensions are N-spin models on a D-dimensional lattice with a toroidal topology, and with a Hamiltonian exhibiting the special structure:

$$H_N = -\sum_s X_s - \sum_c Z_c. \qquad (4.1)$$

Here, $X_s = \otimes_{j \in s} \sigma_j^x$, $Z_c = \otimes_{j \in c} \sigma_j^z$ are products of Pauli matrices belonging to certain finite sets on the lattice called "stars" and "cubes" such that all X_s, Z_c commute. The observables X_s, Z_c generate an abelian subalgebra $\mathcal{A}_{ab}$ in the total algebra $\mathcal{M}_{2^N}$. The commutant of $\mathcal{A}_{ab}$, denoted by $\mathcal{C}$, is noncommutative, and provides a natural basis for encoded qubits. Indeed, similarly to the construction for a 1D Ising model (1.4) one can define *bare qubit observables* X^μ, $Z^\mu \in \mathcal{C}$ where $\mu = 2, 3, 4$ corresponds to D independent encoded qubits. They are products of the corresponding Pauli matrices over topologically nontrivial loops (surfaces). The choice of loops is, of course, non unique.

The Davies generators for the Kitaev models are particularly simple, due to a strict locality of the model (absence of wave propagation). This property implies that the Fourier components in (3.6) are local and correspond to only a few Bohr frequencies, independent of the size of the system.

This makes the analysis of spectral properties of the Davies generator feasible. Despite this simplification the proofs of the results are too involved to be reproduced here; we refer the reader to [10–12] for details, and present here only heuristic arguments.

For the 2D Kitaev model the terms containing σ^x, σ^z in the interaction Hamiltonian (3.2) are sufficient to guarantee all properties (D1)–(D6). The form of the Markovian master equation in the Heisenberg picture is the following

$$
\frac{dA}{dt} = i[H_N, A] + \frac{1}{2} \sum_{j=1}^{N} \left\{ \left(a_j^\dagger [A, a_j] + [a_j^\dagger, A] \, a_j \right. \right.
$$

$$
\left. + e^{-2\beta} a_j [A, a_j^\dagger] + e^{-2\beta} [a_j, A] \, a_j^\dagger \right) - [a_j^0, [a_j^0, A]] \right\}
$$

$$
+ \frac{1}{2} \sum_{j=1}^{N} \left\{ \left(b_j^\dagger [A, b_j] + [b_j^\dagger, A] \, b_j + e^{-2\beta} b_j [A, b_j^\dagger] + e^{-2\beta} [b_j, A] \, b_j^\dagger \right) \right.
$$

$$
\left. - [b_j^0, [b_j^0, X]] \right\}. \tag{4.2}
$$

We do not define here the operators a_j, a_j^0, b_j, b_j^0 but rather give their physical interpretation. The operator a_j ($a_j^\dagger$) annihilates (creates) a pair of excitations (anyons) attached to the site j and corresponding to the part of the Hamiltonian $- \sum Z_c$ in (4.1) (type-Z anyons), while a_j^0 generates diffusion of anyons of the same type. Similarly, the operators b_j, $b_j^\dagger$, b_j^0 act on the type-X anions. It follows that the 2D Kitaev model is equivalent to a gas of noninteracting particles (anyons of two types) which are created/ annihilated in pairs and diffuse. As a consequence there is no mechanism of macroscopic free energy barrier between different phases which could protect even a classical information. Rigorously, it was proved that the the hermitian part of the Davies generator (4.2) possesses a spectral gap independent of the size N and therefore no metastable observables exist. Two main mathematical tools used in the proof are: (1) the fact that for a positive operator K acting on the Hilbert space the inequality $K^2 \geq cK$, $c > 0$ implies that a spectral gap of K is bounded from below by the number c, (2) the Davies generator is a sum of many negatively defined terms, some of them can be skipped to simplify estimations without increasing the spectral gap.

The stability properties of the 4D Kitaev model are much more interesting. Here, the physical picture is rather similar to droplets in the 2D Ising model [14]. The basic excitations of the system are represented by closed

loops with energy proportional to the loops' length providing the mechanism of a macroscopic energy barrier separating topologically nonequivalent
spin configurations (3D model provides this mechanism for one type of
excitations only). The structure of the evolution equation is always similar
to (4.2) with the operators $a_j^\dagger$, $b_j^\dagger$ creating excitations of two types and a_j^0,
b_j^0 changing the shape of excitations but not their energy.

The structure of exponentially stable qubit observables $\tilde{X}^\mu$, $\tilde{Z}^\mu \in \mathcal{C}$
with $\mu = 1, 2, 3, 4$ is similar to the construction for the 1D Ising model (1.5)

$$\tilde{X}^\mu = X^\mu F_x^\mu, \quad \tilde{Z}^\mu = Z^\mu F_z^\mu, \tag{4.3}$$

where F_x^μ, F_z^μ are hermitian elements of the algebra $\mathcal{A}_{ab}$ with eigenvalues
± 1. One should notice that the bare qubit observables X^μ are highly
unstable with relaxation times $\sim \sqrt{N}$. The main tools in the proof of
metastability are: the Peierls argument applied to classical "submodels"
of the 4D-Kitaev model generated either by $-\sum_s X_s$ or $-\sum_c Z_c$, and the
following inequality

$$-\langle A, \mathcal{L}^* A \rangle_\beta \leq 2 \max_{\{\omega\}} \{R(\omega)\} \sum_\alpha \langle [S_\alpha, A], [S_\alpha, A] \rangle_\beta, \tag{4.4}$$

valid for any Davies generator (3.7), (3.8) and any A in the eigenspace of
$[H_N, \cdot]$. The formula (4.4) is very useful because it involves only S_α instead
of Fourier components $S_\alpha(\omega)$.

The metastable observable (say $\tilde{X}^\mu$) is constructed by the following
operational procedure, which determines its outcomes:

1. Perform a measurement of all observables σ_j^x.
2. Compute the value of the bare observable X^μ multiplying the outcomes
   for spins belonging to the "surface" which defines X^μ.
3. Perform a certain classical algorithm (polynomial in N) which allows to
   compute from the σ_j^x- measurement data the value ± 1 of "correction",
   i.e., the eigenvalue of F_x^μ.
4. Multiply the bare value by the correction to get the outcome of $\tilde{X}^\mu$.

The values of qubit observables of above are obtained by the efficient
operational measurement procedure. However, this measurement is highly
destructive because the observables σ_j^x do not commute with the total
Hamiltonian (4.1). Therefore, this measurement cannot be used for the
standard procedure of state initialization, which is necessary to operate
on QM.

5. Concluding Remarks

The models emerging as attempts to design QM, like those proposed by Kitaev, are interesting from the point of view of mathematical physics. Their equilibrium and nonequilibrum properties are nontrivial, nevertheless tractable by rigorous methods. In particular, they give interesting and relatively simple examples of quantum irreversible dynamics governed by Davies generators. New methods of studying their ergodic/spectral properties were developed which should be applicable for other types of models as well. The conclusions which can be drawn from the examples are not optimistic for the very idea of QM. Namely, either no metastable qubit exist or there are serious problems with their accessibility and control. One can think about a kind of "Heisenberg relation" due to the fact that the same physical interactions used to control a system provide its coupling to an environment. Obviously, further rigorous studies of models are necessary to clarify these questions. For example, one cannot exclude that the ultimate bounds on the efficiency of quantum information processing will be provided by phenomenological thermodynamics, in particular by its Second Law [6].

References

1. D. Aharonov and M. Ben-Or, "Fault-tolerant quantum computation with constant error rate", *SIAM J. Comput.* **38** (2008), 1207.
2. E. Knill, R. Laflamme, A. Ashikhmin, H. Barnum, L. Viola and W. Zurek, "Introduction to quantum error correction", *LA Science* **27** (2002), 188.
3. R. Alicki, M. Horodecki, P. Horodecki and R. Horodecki, "Dynamical description of quantum computing: Generic nonlocality of quantum noise", *Phys. Rev.* **A65** (2002), 062101.
4. R. Alicki, D. Lidar and P. Zanardi, "Internal consistency of fault-tolerant quantum error correction in light of rigorous derivations of the quantum Markovian limit", *Phys. Rev.* **A73** (2006), 052311.
5. R. Alicki and M. Horodecki, "Can one build a quantum hard drive? A no-go theorem for storing quantum information in equilibrium systems", arXiv:quant-ph/0603260.
6. R. Alicki, "Quantum memory as a perpetuum mobile of the second kind", arXiv:0901.0811.
7. E. B. Davies, "Markovian master equations", *Commun. Math. Phys.* **39** (1974), 91.
8. G. Lindblad, "On the generators of quantum dynamical semigroups", *Commun. Math. Phys.* **48** (1976), 119.
9. V. Gorini, A. Kossakowski and E. C. G. Sudarshan, "Completely positive dynamical semigroups of N-level systems", *J. Math. Phys.* **17** (1976), 821.

10. R. Alicki, M. Fannes and M. Horodecki, "A statistical mechanics view on Kitaev's proposal for quantum memories", *J. Phys. A: Math. Theor.* **40** (2007), 6451.
11. R. Alicki, M. Fannes and M. Horodecki, "On thermalization in Kitaev's 2D model", *J. Phys. A: Math. Theor.* **42** (2009), 065303.
12. R. Alicki, M. Horodecki, P. Horodecki and R. Horodecki, "On thermal stability of topological qubit in Kitaev's 4D model", arXiv:0811.0033.
13. A. Y. Kitaev, "Fault-tolerant quantum computation by anyons", *Annals Phys.* **303** (2003), 2.
14. E. Dennis, A. Kitaev, A. Landahl and J. Preskill, "Topological quantum memory", *J. Math. Phys.* **43** (2002), 4452.
15. R. Alicki, "Limited thermalization for the Markov mean-field model of N atoms in thermal field", *Physica* **A150** (1988), 455.
16. D. A. Lidar, I. L. Chuang and K. B. Whaley, "Decoherence-free subspaces for quantum computation", *Phys. Rev. Lett.* **81** (1998), 2594.
17. R. Alicki and K. Lendi, *Quantum Dynamical Semigroups and Applications*, II nd. edn. LNP 717, Springer, Berlin, 2007.

TWO MATHEMATICAL PROBLEMS IN
QUANTUM INFORMATION THEORY

Alexander S. Holevo

Steklov Mathematical Institute
Russian Academy of Sciences
Gubkina 8, 119991 Moscow, Russia
E-mail: holevo@mi.ras.ru

We survey the modern status of two major analytical problems in quantum information theory. One is the famous additivity conjecture for information quantities related to the classical capacity of quantum channel. Another is less familiar and less studied problem of optimizers of the information quantities characterizing Bosonic Gaussian channels. We show that for gauge-invariant channels, the validity of the Gaussian minimizer conjecture for the output entropy implies Gaussianity of the ensemble maximizing the χ-capacity.

1. Formulation

1.1. *The additivity problem*

For a quantum channel Φ (see Sec. 2.1), a noncommutative analog of the Shannon capacity, which we call χ-*capacity*, is defined by

$$C_\chi(\Phi) = \sup_{\{\pi_j, \rho_j\}} \left(\left(\sum_j \pi_j \Phi[\rho_j] \right) - \sum_j \pi_j H(\Phi[\rho_j]) \right), \qquad (1.1)$$

where the supremum is over all quantum ensembles, that is finite collections of states (density operators) $\{\rho_1, \ldots, \rho_n\}$ with corresponding probabilities $\{\pi_1, \ldots, \pi_n\}$. Here $H(\rho) = -\operatorname{Tr} \rho \log \rho$ denotes the von Neumann entropy of the density operator ρ. The quantity (1.1) is closely related to the capacity of quantum channel Φ for transmitting classical information [20].

The problem is: does the *the additivity property*

$$C_\chi(\Phi_1 \otimes \Phi_2) = C_\chi(\Phi_1) + C_\chi(\Phi_2) \qquad (1.2)$$

109

hold for tensor product of any pair of quantum channels Φ_1, Φ_2? The problem can be traced back to [3], see also [20].

Such additivity is proved rather simply for the Shannon capacity of parallel classical channels, but in the quantum case it is established only for few cases including both "very classical" and "very quantum" channels. Similar additivity conjectures exist also for other interesting entropy characteristics of quantum channels such as the *minimal output entropy*

$$\check{H}(\Phi) = \inf_\rho H(\Phi[\rho]), \tag{1.3}$$

namely

$$\check{H}(\Phi_1 \otimes \Phi_2) \overset{?}{=} \check{H}(\Phi_1) + \check{H}(\Phi_2). \tag{1.4}$$

In the case $\dim \mathcal{H} < \infty$, which we assume in the first part of this paper, both inf in (1.3) and sup in (1.1) are attained by continuity-and-compactness argument and will be replaced by min and max correspondingly. However it is not so in infinite-dimensional case, and attainability of the extrema requires separate study, see Sec. 2.4.

The reason for the additivity to hold in the classical case is most easily seen for the minimal output entropy. Since entropy is concave, it attains its minimum at an extreme point of the simplex $\mathcal{P}(\mathcal{X})$ of classical states, i.e. probability distributions on an underlying phase space $\mathcal{X}$, which is just a distribution degenerated at some point $x \in \mathcal{X}$. But every extreme point of $\mathcal{P}(\mathcal{X}_1 \times \mathcal{X}_2)$ is product of extreme points of $\mathcal{P}(\mathcal{X}_j)$:

$$\text{ext}\,\mathcal{P}(\mathcal{X}_1 \times \mathcal{X}_2) = \text{ext}\,\mathcal{P}(\mathcal{X}_1) \times \text{ext}\,\mathcal{P}(\mathcal{X}_2), \tag{1.5}$$

which immediately implies the additivity (1.4) in the classical case.

A detailed account of the problem status until 2006 was given in the author's talk at the last ICM [23]. A year later came important findings of Winter [49] and Hayden [14] which showed existence of a pair of channels breaking the additivity of closely related quantity – the minimal output Rényi entropy for all values of the parameter $p > 1$. Quite recently, basing on this progress, Hastings [12] announced proof of existence of channels breaking the additivity conjecture (1.4) corresponding to $p = 1$, in very high dimensions. This still leaves open the question of explicit demonstration of a counterexample; other important questions concern what happens in small and moderate dimensions (e.g. additivity for non-unital qubit channels) and additivity for certain important classes like Bosonic Gaussian channels.

1.2. *The problem of Gaussian optimizers*

Particularly interesting class in infinite dimensional case constitute quantum Gaussian channels (see Sec. 2.3), which can be considered as a natural generalization of Gaussian channels in classical information theory, where it is well known that their Shannon capacity under quadratic (power) constraint is attained on Gaussian inputs. The conjecture is then that also *quantum Gaussian channels have Gaussian optimizers*, e.g. the χ-capacity of a quantum Gaussian channel Φ under additional input constraint of the type $\text{Tr}\left(\sum_j \pi_j \rho_j\right) F \leq E$, where F is a "quadratic Hamiltonian", is attained on a (continuous) Gaussian ensemble of pure Gaussian states. Or even more simply, the unconditional minimum of the output entropy is attained on a pure quantum Gaussian state.

In the commutative analysis, a related problem of Gaussian maximizers which refers to L_p-norms of integral operators with Gaussian kernel has been studied rather exhaustively (see e.g. the paper of Lieb [32] and references therein).

There is an intriguing connection with the additivity problem. In the paper [50] it is shown, by using the quantum central limit theorem, that if the additivity holds, then the average state of the optimal ensemble can be chosen Gaussian; however this still leaves untouched the hard problem of Gaussian optimizer for the second term in χ-capacity which turns out to be intrinsically connected to another key quantity — entanglement of formation. Thus, although the Gaussian optimizers problem looks more special, it might be even more difficult to solve than the additivity conjecture.

Another connection with the additivity conjecture comes when one intentionally restricts to Gaussian states, respectively ensembles, in the optimization problems (1.3), respectively (2.15) for a Gaussian channel Φ, which results in the quantities $\check{H}^{Gauss}(\Phi)$, respectively $C_\chi^{Gauss}(\Phi)$. One then asks whether the additivity properties (1.4), respectively (2.18) hold for these Gaussian quantities. Surprisingly, even this substantially simplified problem has so far only partial solution: the additivity was proved only for some special classes of quantum Gaussian channels [41, 18].

2. Quantum Channels

2.1. *Additivity and entanglement*

Let $\mathcal{H}$ be a finite-dimensional Hilbert space of dimensionality d and $\mathfrak{L}(\mathcal{H})$ be the algebra of all linear operators in $\mathcal{H}$.

Finite quantum system is described by the noncommutative algebra $\mathfrak{L}(\mathcal{H})$. The convex subset of $\mathfrak{L}(\mathcal{H})$

$$\mathfrak{S}(\mathcal{H}) = \{\rho : \rho^* = \rho \geq 0, \operatorname{Tr}\rho = 1\}$$

is called the *quantum state space*. Operators ρ from $\mathfrak{S}(\mathcal{H})$ are called *density operators* or *quantum states*. The state space is a convex set with the extreme boundary

$$\mathfrak{P}(\mathcal{H}) = \operatorname{ext}\mathfrak{S}(\mathcal{H}) = \{\rho : \rho \geq 0, \operatorname{Tr}\rho = 1, \rho^2 = \rho\}.$$

Thus extreme points of $\mathfrak{S}(\mathcal{H})$, which are also called *pure states*, are one-dimensional projectors, $\rho = P_\psi$ for a vector $\psi \in \mathcal{H}$ with unit norm, see, e.g. [20].

Classical system can be described by the commutative subalgebra $\mathfrak{C}(\mathcal{H})$, consisting of all operators diagonal in a fixed orthonormal basis. Classical states are the density matrices $\rho \in \mathfrak{C}(\mathcal{H})$, which have form $\rho = \operatorname{diag}[\pi(x)]_{x=1,\ldots,d}$ with $\pi = \{\pi(x)\}$ – a probability distribution.

We shall consider linear maps Φ which take operators A in d-dimensional unitary space $\mathcal{H}$ to operators $A' = \Phi[A]$ in d'-dimensional $\mathcal{H}'$. The map $\Phi : \mathfrak{L}(\mathcal{H}) \to \mathfrak{L}(\mathcal{H}')$ is called *positive* if $A \geq 0$ implies $\Phi[A] \geq 0$.

The dual Φ^* of the map Φ is uniquely defined by the relation

$$\operatorname{Tr}\Phi[A]B = \operatorname{Tr}A\Phi^*[B]; \quad A, B \in \mathfrak{L}(\mathcal{H}). \tag{2.1}$$

A positive map Φ is called trace preserving if it takes quantum states into quantum states (possibly in another space $\mathcal{H}'$), and *unital* if $\Phi(I) = I'$, where I, I' denote unit operator in the corresponding Hilbert space.

Especially important for us will be the class of completely positive (CP) maps [46, 5]. The map $\Phi : \mathfrak{L}(\mathcal{H}) \to \mathfrak{L}(\mathcal{H}')$ is *completely positive*, if for $d = 1, 2, \ldots$ the maps $\Phi \otimes \operatorname{Id}_d$ are all positive, where $\operatorname{Id}_d : \mathfrak{L}_d \to \mathfrak{L}_d$ is the identity map of the algebra $\mathfrak{L}_d$ of all complex $d \times d$-matrices. It follows that tensor product of CP maps is again CP, since

$$\Phi_1 \otimes \Phi_2 = (\operatorname{Id}_{d'_1} \otimes \Phi_2) \circ (\Phi_1 \otimes \operatorname{Id}_{d_2}).$$

There are positive maps that are not CP, a basic example provided by matrix transposition $A \to A^T$ in a fixed basis.

A completely positive trace preserving map Φ is called *channel*; the *dual channel* Φ^* is a completely positive unital map.

For a channel Φ consider the quantity $\check{H}(\Phi)$ defined in (1.3). Unlike the classical case, there is no obvious reason for the additivity (1.4), because there is no analog of (1.5). In fact

$$\operatorname{ext}\mathfrak{S}(\mathcal{H}_1 \otimes \mathcal{H}_2) \supsetneqq \operatorname{ext}\mathfrak{S}(\mathcal{H}_1) \times \operatorname{ext}\mathfrak{S}(\mathcal{H}_2), \tag{2.2}$$

since apparently, there are continually many pure states P_ψ in $\mathcal{H}_1 \otimes \mathcal{H}_2$, given by vectors ψ not representable as a tensor product $\psi_1 \otimes \psi_2$ (namely, all linear combinations of such vectors which do not reduce to products). In quantum theory tensor product $\mathcal{H}_1 \otimes \mathcal{H}_2$ describes composite (bipartite) system. Vectors that are not of the form $\psi_1 \otimes \psi_2$, as well as the corresponding pure states, are called *entangled*. In an entangled pure state of a bipartite quantum system, neither of the parts is in a pure state, in sharp contrast to the classical systems.

Turning to the χ-capacity (1.1), the additivity problem can be formulated in physical terms as: Can entanglement between input states increase the classical capacity of quantum channel? The latter is defined as the maximal transmission rate per use of the channel, with coding and decoding chosen for increasing number n of independent uses of the channel

$$\Phi^{\otimes n} = \underbrace{\Phi \otimes \cdots \otimes \Phi}_{n}$$

such that the error probability goes to zero as $n \to \infty$ (see [20]). A basic result of quantum information theory says that such defined capacity $C(\Phi)$ is related to $C_\chi(\Phi)$ by the formula

$$C(\Phi) = \lim_{n \to \infty} (1/n) C_\chi(\Phi^{\otimes n}) = \sup_{n} (1/n) C_\chi(\Phi^{\otimes n}).$$

Since $C_\chi(\Phi)$ is easily seen to be superadditive (i.e., $C_\chi(\Phi_1 \otimes \Phi_2) \geq C_\chi(\Phi_1) + C_\chi(\Phi_2)$), one has $C(\Phi) \geq C_\chi(\Phi)$. If the additivity (1.2) holds for given channel $\Phi_1 = \Phi$ and arbitrary channel Φ_2, then

$$C_\chi(\Phi^{\otimes n}) = n C_\chi(\Phi), \tag{2.3}$$

implying

$$C(\Phi) = C_\chi(\Phi). \tag{2.4}$$

Such a relation is very much welcome mathematically giving a relatively easily computable "single-letter" expression for the classical capacity of the quantum channel Φ. One might ask if formally weaker additivity property (2.3) holds globally, i.e. for all channels. However, a result in [9] shows that this would imply the additivity for all pairs of different channels (1.2). Since this is not the case, the asymptotic equality (2.4) cannot hold for all channels.

2.2. Different forms of the additivity property

From the definition of χ-capacity it follows that

$$C_\chi(\Phi) \leq \max_{\rho \in \mathfrak{S}(\mathcal{H})} H(\Phi(\rho)) - \min_{\rho \in \mathfrak{S}(\mathcal{H})} H(\Phi(\rho)). \qquad (2.5)$$

To find out the intrinsic connection between the output entropy $H(\Phi(\rho))$ and the χ-capacity, let us rewrite the expression (1.1) in the form

$$C_\chi(\Phi) = \max_{\rho \in \mathfrak{S}(\mathcal{H})} [H(\Phi(\rho)) - \hat{H}_\Phi(\rho)], \qquad (2.6)$$

where

$$\hat{H}_\Phi(\rho) = \min_{\pi : \sum_x \pi(x)\rho(x) = \rho} \sum_x \pi(x) H(\Phi(\rho(x)))$$

is the *convex closure* [34] of the output entropy $H(\Phi(\rho))$. The minimum here is taken over arbitrary finite probability distributions on $\mathfrak{S}(\mathcal{H})$. The conjectured superadditivity property is: *for arbitrary state* $\rho_{12} \in \mathfrak{S}(\mathcal{H}_1 \otimes \mathcal{H}_2)$ *and arbitrary channels* Φ_1, Φ_2

$$\hat{H}_{\Phi_1 \otimes \Phi_2}(\rho_{12}) \overset{?}{\geq} \hat{H}_{\Phi_1}(\rho_1) + \hat{H}_{\Phi_2}(\rho_2), \qquad (2.7)$$

where ρ_1, ρ_2 *are the partial traces of* ρ_{12} *in* $\mathcal{H}_1, \mathcal{H}_2$ (for the notion of partial trace see the next section). The function $\hat{H}_\Phi(\rho)$ is a natural generalization of another important quantity in quantum information theory — *entanglement of formation* E_F. In fact

$$\hat{H}_\Phi(\rho) = E_F(V\rho V^*), \qquad (2.8)$$

where V is the Stinespring isometry for the channel Φ [38] and reduces to it when the channel Φ is simply a partial trace.

Proposition 1. *For given channels* Φ_1, Φ_2 *the superadditivity property* (2.7) *implies both additivity properties* (1.2) *and* (1.4).

Proof. Indeed, let ρ_{12}^0 be a minimizer for $H(\Phi_1 \otimes \Phi_2)(\rho_{12})$, then

$$\check{H}(\Phi_1 \otimes \Phi_2) = H((\Phi_1 \otimes \Phi_2)(\rho_{12}^0)) \geq \hat{H}_{\Phi_1 \otimes \Phi_2}(\rho_{12}^0)$$
$$\geq \hat{H}_{\Phi_1}(\rho_1^0) + \hat{H}_{\Phi_2}(\rho_2^0) \geq \check{H}(\Phi_1) + \check{H}(\Phi_2),$$

whence (1.4) follows. On the other hand, (2.7) and subadditivity of quantum entropy

$$H(\sigma_{12}) \leq H(\sigma_1) + H(\sigma_2),$$

imply

$$H((\Phi_1 \otimes \Phi_2)(\rho_{12})) - \hat{H}_{\Phi_1 \otimes \Phi_2}(\rho_{12})$$
$$\leq H((\Phi_1 \otimes \Phi_2)(\rho_{12})) - \hat{H}_{\Phi_1}(\rho_1) - \hat{H}_{\Phi_2}(\rho_2)$$
$$\leq \left[H(\Phi_1(\rho_1)) - \hat{H}_{\Phi_1}(\rho_1) \right] + \left[H(\Phi_2(\rho_2)) - \hat{H}_{\Phi_2}(\rho_2) \right].$$

By using (2.6), we get

$$C_\chi(\Phi_1 \otimes \Phi_2) \leq C_\chi(\Phi_1) + C_\chi(\Phi_2),$$

i.e. (1.2). □

In [24] several individually equivalent formulations of the additivity property for channels with constrained inputs, which formally is substantially stronger than additivity of the unconstrained C_χ, were given. It was shown that the additivity for channels with constrained inputs holds true for certain nontrivial classes of channels, e.g. a direct sum mixture of the identity channel and entanglement breaking or diagonal channel.

Quite remarkably, however, all the additivity properties turn out to be the same *globally*. By combining the correspondence (2.8) and the convex duality technique of [2] with a powerful channel extension technique, which allows to use effectively arbitrariness of channels in question, Shor [45] had shown equivalence of the global properties of additivity of the minimal output entropy, C_χ, E_F and of superadditivity of E_F.

Theorem 1 ([45]). *The conjectures* (1.2), (1.4) *and* (2.7) *are globally equivalent in the sense that if one of them holds true for all channels* Φ_1, Φ_2, *then any other is also true for all channels.*

The channel extension technique was used in [24] to show that additivity for two fixed constrained channels can be reduced to the same problem for some unconstrained channels, and hence, the global additivity for channels with arbitrary input constraints is equivalent to the global additivity without constraints.

These global equivalences also guarantee existence of channels breaking all other forms of the additivity conjecture, if only one of them, say, the minimal output entropy is shown to be nonadditive globally. However, the applicability of individual statements like Proposition 1 remains unaffected.

2.3. *Nonadditivity of quantum entropy quantities*

The quantum Rényi entropy of order $p \geq 0, p \neq 1$ of a density operator ρ is defined as

$$R_p(\rho) = \frac{1}{1-p} \log \operatorname{Tr} \rho^p, \qquad (2.9)$$

and in the limit $p \to 1$ the quantum Rényi entropies uniformly converge to the von Neumann entropy of a density operator ρ

$$\lim_{p \to 1} R_p(\rho) = -\operatorname{Tr} \rho \log \rho \equiv H(\rho).$$

Defining the minimal output Rényi entropy of the channel Φ

$$\check{R}_p(\Phi) = \min_{\rho \in \mathfrak{S}(\mathcal{H})} R_p(\Phi(\rho)),$$

one has the additivity conjecture similar to (1.4)

$$\check{R}_p(\Phi_1 \otimes \Phi_2) \overset{?}{=} \check{R}_p(\Phi_1) + \check{R}_p(\Phi_2). \tag{2.10}$$

Again, the inequality $\leq$ is obvious here. Note, that validity of (2.10) for some specific channels Φ_1, Φ_2 and p close to 1 implies (1.4) for these Φ_1, Φ_2.

There are several important classes of channels for which the property (2.10) can be proved for all p, including all entanglement-breaking channels (Shor [44], King [28]), all unital qubit channels (King [29]) and the depolarizing channel (King [30]) as well as for complementary channels (Holevo [22], King $et\ al.$ [31]). A significant role in the proofs is played by the Lieb-Thirring inequality [33]

$$\operatorname{Tr}(AB)^p \leq \operatorname{Tr} A^p B^p, \tag{2.11}$$

for $A, B \in \mathfrak{L}(\mathcal{H})$, $A, B \geq 0$, and $p \geq 1$.

However, there is an explicit example of $transpose\text{-}depolarizing\ channel$ [48]), where the additivity breaks for $d = \dim \mathcal{H} \geq 3$ and large enough p:

$$\Phi(\rho) = \frac{1}{d-1} \left[I - \rho^T \right].$$

In particular, (2.10) with $\Phi_1 = \Phi_2 = \Phi$ fails to hold for $p \geq 4,7823$ if $d = 3$. Nevertheless, the additivity of $\check{H}(\Phi)$ and of $C_\chi(\Phi)$ holds for this channel, as shown in [37, 7, 1].

A breakthrough in the negative solution of the conjecture (2.10) came in 2007. It was shown that in very high dimensions there always exist channels for which (2.10) does not hold for any $p > 2$ ([49]) and for any $1 < p < 2$ ([14]). Specifically, in [49] it was shown that the additivity of the Rényi entropy with $p > 2$ for the couple of channels $\Phi, \bar{\Phi}$ (complex conjugate in a fixed basis) breaks with probability tending to 1 with $d \to \infty$ for the uniform mixture of unitary evolutions

$$\Phi(\rho) = \frac{1}{n} \sum_{j=1}^n U_j \rho U_j^*, \tag{2.12}$$

where $U_j; j = 1, \ldots, n$ is a sequence of random independent unitary operators distributed according to the normalized Haar measure, and $n = \frac{134}{\varepsilon^2} d \log d$. A simple but efficient upper bound for $\check{R}_p(\Phi \otimes \bar{\Phi})$ is

$$\check{R}_p(\Phi \otimes \bar{\Phi}) \leq R_p((\Phi \otimes \bar{\Phi})(|\Psi\rangle\langle\Psi|)) \leq \frac{p}{p-1} \log n, \qquad (2.13)$$

where $|\Psi\rangle$ is the maximally entangled vector in the same basis, and in the second inequality one uses the special property $(U \otimes \bar{U})|\Psi\rangle = |\Psi\rangle$ for arbitrary unitary U.

More difficult is the probabilistic lower bound

$$\lim_{d \to \infty} \mathsf{P}\left\{ \check{R}_p(\Phi) \geq \log \frac{d}{\varepsilon} \right\} = 1$$

for random Φ of the form (2.12). A key ingredient is a large deviation estimate for sums of random operators inspired by the classical Bernstein-Chernoff-Hoeffding inequality [15], which show that (2.12) becomes close to completely depolarizing channel in the sense of operator norm

$$\left\| \Phi(\rho) - \frac{I}{d} \right\|_\infty \leq \frac{\varepsilon}{d},$$

while preserving entanglement to the degree given by (2.13).

Hayden [14] used a general open system representation of the channel Φ with the the random evolution operator distributed uniformly over the group of unitaries. Then the upper bound

$$\check{R}_p(\Phi \otimes \bar{\Phi}) \leq R_p((\Phi \otimes \bar{\Phi})(|\Psi\rangle\langle\Psi|)) \leq p \log d + O(1)$$

sufficient if $1 < p < 2$,[a] can be obtained similarly to (2.13), while the probabilistic estimate

$$\lim_{d \to \infty} \mathsf{P}\left\{ \check{R}_p(\Phi) \geq \log d - 2 \right\} = 1$$

is based on another large deviation phenomenon — the measure concentration which is used to establish concentration of the output entropy [16]. This is closely related to the early important observations going back to Lubkin [36] that given a random uniformly distributed pure state vector $|\psi_{AB}\rangle$ of a composite system AB, the entropy of the partial state ρ_B of the smaller system tends to its maximal value $\log d_B$ as $\log d_A \to \infty$, and hence ρ_B becomes almost chaotic.

---

[a]The whole range $1 < p$ can be also covered by this construction as shown later by Winter. Subsequently, there were also negative results for $p = 0$ and $p \approx 0$ [6]. For the later improvements see the merged article by Hayden and Winter [17].

Hastings [12] explored this phenomenon at full strength along with exact probability distribution of the spectrum of ρ_B found in [35, 51]. He generalized Winter's construction by considering non-uniform mixtures of unitary evolutions with random weights distributed in such a way that the output of the channel Φ is distributed precisely as ρ_A from the uniformly distributed $|\psi_{AB}\rangle$, while the output of the complementary channel $\tilde{\Phi}$ – as ρ_B. He uses the estimate

$$\check{H}(\tilde{\Phi} \otimes \bar{\tilde{\Phi}}) \leq H\big((\tilde{\Phi} \otimes \bar{\tilde{\Phi}})(|\Psi\rangle\langle\Psi|)\big) \leq 2\log n - \frac{\log n}{n}$$

again similar to (2.13), while the most difficult part

$$\lim_{d \to \infty} \mathsf{P}\left\{\check{H}(\tilde{\Phi}) > \log n - \frac{\log n}{2n}\right\} > 0$$

for n, d/n large enough is obtained by comparing the consequences of the aforementioned distribution of the spectrum of ρ_B and an original probability estimate of the minimal output entropy (Lemma 4 of [12]), avoiding straightforward use of ε-nets and the "union bound", which does not attain the goal in the case $p = 1$. Hastings gave only a sketch, and the detailed proof following Hastings' approach was given by Fukuda, King and Moser [8].

Although, combined with theorem 1 this gives a definite answer to the additivity conjecture, several important issues remain open. All the proofs above use the technique of random unitary operators or random states and as such are not constructive: they provide only evidence for existence of counterexamples but do not allow to actually produce them. Attempts to give estimates for the dimensions in which nonadditivity can happen so far has led to overwhelmingly high values: the detailed estimates made in [8] gave $n \approx 3.9 \times 10^4$, $d \approx 7.8 \times 10^{32}$ breaking the additivity by the quantity of the order 10^{-5}. While this does not exclude possibility of better estimates, based perhaps on a different (but yet unknown) approach, it casts doubt to finding concrete counterexamples by computer simulation of random unitary channels. It remains a mystery what happens in realistic dimensions: perhaps the additivity still holds generically for some unknown reason, or its violation is so tiny that it cannot be catched by numerical simulations.[b]

---

[b]That such a possibility is not excluded in quantum information problems is well illustrated by Shor's counterexample to Levitin's conjecture concerning maximizers for accessible information [43].

From the physical side, that would mean that, strangely, entangled encodings are rather useless "practically", in contrast to entangled decodings which show the superadditivity starting from $d = 2$ [20].

In any case, this work is fundamental from the mathematical side, finally closing attempts of a general proof and stressing the importance of continuing efforts to find further particular cases where the additivity holds for this or another reason.

2.4. *Infinite-dimensional channels*

The additivity problem is still open for the minimal dimension 2: it is not known if the additivity holds for all nonunital qubit channels, although a strong numerical evidence in favor of this was given in [39, 13]. Nevertheless there are several good reasons to consider the problem in infinite dimensions. There is a good chance that additivity holds for important and interesting class of Gaussian channels that act in infinite dimensional Hilbert space, see e.g. [11, 10].

Shor's channel extension used in the proof of equivalence of different forms of the global additivity conjecture for finite dimensional channels is related to weird discontinuity of the χ-capacity as a function of channel in infinite dimensions. This also calls for a mathematically rigorous treatment of the entropic quantities related to the classical capacity of infinite dimensional channels [25].

In infinite dimensions, analysis of continuity properties of the entropic characteristics of an infinite-dimensional channel becomes important since, as it is well known, the entropy may have rather pathological behavior. It is lower semicontinuous and "almost everywhere" infinite [47]. Another issue is the study of conditions for compactness of subsets of quantum states and ensembles, giving a key for attainability of extrema in expressions for the capacity and the convex closure of the output entropy. Such a study was undertaken in series of works [21, 25, 42].

There are two important features essential for channels in infinite dimensions. One is the necessity of the input constraints (such as mean energy constraint for Gaussian channels) to prevent from infinite capacities (although considering input constraints was shown quite useful also in the study of the additivity conjecture for channels in finite dimensions [24]). Another is the natural appearance of infinite, and, in general, "continuous" state ensembles understood as probability measures on the set of all quantum states. By using compactness criteria from probability theory and operator theory one can show that the set of all such generalized ensembles with the barycenter in a compact set of states is itself weakly compact. With this at hand a sufficient condition for existence of an optimal generalized ensemble for a constrained quantum channel can be given. This condition can be efficiently verified in the case of Bosonic Gaussian channels with constrained mean energy [25].

The *generalized ensemble* is defined as a Borel probability measure π on the state space $\mathfrak{S}(\mathcal{H})$. The average state of the ensemble π is given by the baricenter

$$\bar{\rho}_\pi = \int_{\mathfrak{S}(\mathcal{H})} \rho\pi(d\rho).$$

Let F be positive self-adjoint operator usually representing energy, E — a positive constant. Then the constrained χ-capacity of channel Φ is defined as

$$C_\chi(\Phi, F, E) = \sup_{\pi:\operatorname{Tr}\bar{\rho}_\pi F \leq E} \left[H(\Phi(\bar{\rho}_\pi)) - \int_{\mathfrak{S}(\mathcal{H})} H(\Phi(\rho))\pi(d\rho) \right] \quad (2.14)$$

$$= \sup_{\rho:\operatorname{Tr}\rho F \leq E} [H(\Phi[\rho]) - \hat{H}_\Phi(\rho)], \quad (2.15)$$

where

$$\hat{H}_\Phi(\rho) = \inf_{\pi:\bar{\rho}_\pi = \rho} \int_{\mathfrak{S}(\mathcal{H})} H(\Phi(\sigma))\pi(d\sigma). \quad (2.16)$$

Similarly to (1.1), there is a relation to the constrained classical capacity

$$C(\Phi, F, E) = \lim_{n \to \infty} \frac{1}{n} C_\chi(\Phi^{\otimes n}, F^{(n)}, nE). \quad (2.17)$$

If the additivity property

$$C_\chi(\Phi^{\otimes n}, F^{(n)}, nE) = nC_\chi(\Phi, F, E), \quad (2.18)$$

holds for the channel Φ, then

$$C(\Phi, F, E) = C_\chi(\Phi, F, E).$$

This is closely related to superadditivity of the convex closure of the output entropy (2.7), which in particular implies additivity of the χ-capacity with linear constraints [24].

3. Gaussian Systems

3.1. *Canonical commutation relations, symplectic space and complex structures*

In quantum mechanics the canonical commutation relations (CCR) arise in quantization either of a mechanical system with finite degrees of freedom or of a classical field by representing it as an infinite collection of oscillators and hence as a mechanical system with infinite number degrees of freedom. However, in quantum optics one usually deals only with a finite number of relevant oscillator frequencies, thus again reducing to mechanical system with a finite number s of degrees of freedom. We give below a very brief account of the CCR and Bosonic Gaussian systems based on them, see e.g. [19, 26, 4] for further detail.

Such a system is formally described by *canonical observables* $q_j, p_j; j = 1, \ldots, s$, which are operators in underlying Hilbert space satisfying the Heisenberg CCR

$$[q_j, p_k] = i\delta_{jk}I, \quad [q_j, q_k] = 0, \quad [p_j, p_k] = 0 \tag{3.1}$$

Let us introduce the column vector of operators

$$R = [q_1, p_1, \ldots, q_s, p_s]^\top,$$

and the real column $2s$-vector $z = [x_1, y_1, \ldots, x_s, y_s]^\top$ so that

$$R^\top z = \sum_{j=1}^{s}(x_j q_j + y_j p_j).$$

Mathematically it is convenient to approach to the CCR via the unitary operators

$$W(z) = \exp i\, R^\top z \tag{3.2}$$

satisfying the *Weyl-Segal CCR*

$$W(z)W(z') = \exp\left[\frac{i}{2}\Delta(z, z')\right]W(z + z'), \tag{3.3}$$

where

$$\Delta(z, z') = \sum_{j=1}^{s}(x'_j y_j - x_j y'_j) = z^\top \Delta z' \tag{3.4}$$

is the canonical *symplectic form*, where $\Delta = [\Delta_{jk}]$ is the $(2s) \times (2s)$-skew-symmetric *commutation matrix* of components of the vector R,

$$\Delta = \begin{bmatrix} 0 & -1 & & & \\ 1 & 0 & & & \\ & & \ddots & & \\ & & & 0 & -1 \\ & & & 1 & 0 \end{bmatrix} \equiv \mathrm{diag} \begin{bmatrix} 0 & -1 \\ 1 & 0 \end{bmatrix}. \tag{3.5}$$

The space Z of real $2s$-vectors equipped with a nondegenerate skew-symmetric form $\Delta(z, z')$ is what one calls a *symplectic space*. It represents the phase space of the classical system, the quantum version of which is described by CCR.

Lemma 1. *Let $\alpha(z, z') = z^\top \alpha z'$ be an inner product in the symplectic space (Z, Δ). Then there is a basis $\{e_j, h_j; j = 1, \dots, s\}$ in Z in which the forms Δ, α have the matrices*

$$\tilde{\Delta} = \mathrm{diag} \begin{bmatrix} 0 & -1 \\ 1 & 0 \end{bmatrix}; \quad \tilde{\alpha} = \mathrm{diag} \begin{bmatrix} \alpha_j & 0 \\ 0 & \alpha_j \end{bmatrix}, \tag{3.6}$$

with $\alpha_j > 0$.

Indeed, consider the matrix $\hat{\alpha} = \Delta^{-1} \alpha$ which is the matrix of the operator (denoted by the same symbol) satisfying

$$\alpha(z, z') = \Delta(z, \hat{\alpha} z').$$

The operator $\hat{\alpha}$ is skew-symmetric in the Euclidean space (Z, α) : $\hat{\alpha}^* = -\hat{\alpha}$, hence there exists an orthogonal basis $\{e_j, h_j\}$ in (Z, α) and positive numbers $\{\alpha_j\}$ such that

$$\hat{\alpha} e_j = -\alpha_j h_j; \quad \hat{\alpha} h_j = \alpha_j e_j.$$

Choosing the normalization $\alpha(e_j, e_j) = \alpha(h_j, h_j) = \alpha_j$ gives the basis with the required properties.

Any basis in which the matrix of $\Delta(z, z')$ has the canonical form (3.5) is called *symplectic*. Denoting by T the transition matrix from the initial basis in Z to the symplectic basis, i.e. the matrix with the columns $\{e_j, h_j; j = 1, \dots, s\}$ we have

$$\tilde{\Delta} = T^\top \Delta T; \quad \tilde{\alpha} = T^\top \alpha T.$$

We can also define the canonical variables $\tilde{q}_j = R^\top e_j, \tilde{p}_j = R^\top h_j; j = 1, \dots, s$, so that $T^\top R = \tilde{R}$, where $\tilde{R} = [\tilde{q}_1, \tilde{p}_1, \dots, \tilde{q}_s, \tilde{p}_s]^\top$.

Operator J in Z is called *operator of complex structure* if

$$J^2 = -E, \tag{3.7}$$

where E is the identity operator in Z, and

$$\Delta(z, Jz) \geq 0; \quad z \in Z. \tag{3.8}$$

or in the matrix notations

$$-\Delta J = J^\top \Delta \geq 0. \tag{3.9}$$

For any inner product α on Z defined by the corresponding positive-definite matrix, there is an operator of complex structure J commuting with the operator $\hat{\alpha} = \Delta^{-1}\alpha$:

$$[J, \Delta^{-1}\alpha] = 0, \tag{3.10}$$

namely, the orthogonal operator J from the polar decomposition $\hat{\alpha} = J|\hat{\alpha}| = |\hat{\alpha}|J$ of the skew-symmetric operator $\hat{\alpha}$. One has $J = T\tilde{J}T^{-1}$ where $\tilde{J}$ has the matrix

$$\tilde{J} = -\tilde{\Delta} = \mathrm{diag} \begin{bmatrix} 0 & 1 \\ -1 & 0 \end{bmatrix} \tag{3.11}$$

in the symplectic basis associated with the inner product $\alpha(z, z')$.

With every complex structure one can associate the cyclic one-parameter group $\{e^{\varphi J}\}$ of symplectic transformations which we call the *gauge group*. The gauge group in Z induces the unitary group of the *gauge transformations* in $\mathcal{H}$ according to the formula

$$W(e^{\varphi J}z) = e^{-i\varphi G}W(z)e^{i\varphi G}, \tag{3.12}$$

where $G = \frac{1}{2}R^\top J\Delta^{-1}R$ is selfadjoint positive operator in $\mathcal{H}$. In terms of generators, this reduces to

$$R^\top e^{\varphi J} = e^{-i\varphi G}R^\top e^{i\varphi G}. \tag{3.13}$$

A selfadjoint operator F in $\mathcal{H}$ is called *gauge invariant* if

$$e^{-i\varphi G}Fe^{i\varphi G} = F$$

for all real φ. Consider the quadratic operator $F = R^\top \epsilon R$, where ϵ is a symmetric positive-definite (energy) matrix. By using (3.13) we find that it is gauge invariant if $e^{\varphi J}\epsilon\, e^{\varphi J^\top} = \epsilon$, or, equivalently, $J\epsilon + \epsilon J^\top = 0$, which by (3.9) is the same as

$$[J, \epsilon\Delta] = 0.$$

For every energy matrix ϵ, there is an operator of complex structure satisfying this condition: it is the orthogonal operator from the polar decomposition of $\epsilon\Delta$. One has $J = T\tilde{J}T^{-1}$ where $\tilde{J}$ has the matrix (3.11) in the symplectic basis associated with the inner product $\epsilon(z, z') = \Delta(z, \epsilon\Delta z') = z^{\top}\Delta\epsilon\Delta z'$.

In quantum optics, where the field is reduced to a finite collection of oscillator modes, there is a preferred complex structure arising from the oscillator Hamiltonian which gives rise to creation-annihilation operators and amounts then to usual multiplication by i.

3.2. Gaussian states and channels

Let $W(z); z \in Z$, be a representation of the CCR in $\mathcal{H}$. The state ρ is called *Gaussian*, if its quantum characteristic function

$$\phi(z) = \text{Tr}\,\rho W(z)$$

has the form

$$\phi(z) = \exp\left(i\,m^{\top}z - \frac{1}{2}z^{\top}\alpha z\right), \tag{3.14}$$

where m is a column $(2s)$-vector and α is a real symmetric $(2s)\times(2s)$-matrix satisfying the *matrix uncertainty relation*

$$\alpha - \frac{i}{2}\Delta \geq 0. \tag{3.15}$$

Let J be an operator of complex structure in Z and let $\{e^{i\varphi G}\}$ be the corresponding gauge group in $\mathcal{H}$. From (3.12) it follows that the Gaussian density operator ρ is gauge invariant, $e^{-i\varphi G}\rho\,e^{i\varphi G} = \rho$, $\varphi \in \mathbf{R}$, if and only if its characteristic function satisfies $\phi(e^{\varphi J}z) = \phi(z)$, which is equivalent to $m = 0$ and $J^{\top}\alpha + \alpha J = 0$. By using (3.9), the last condition can be written as (3.10). As we have seen in Sec. 3.1, for arbitrary α there is at least one operator of complex structure in Z satisfying this condition, namely the orthogonal operator J from the polar decomposition of operator $\Delta^{-1}\alpha$.

The condition (3.15) amounts to the fact that in the diagonal form (3.6)

$$\alpha_j \geq \frac{1}{2}, \quad j = 1, \ldots, s. \tag{3.16}$$

When $\alpha_j = \frac{1}{2}$, the Gaussian state is pure and has the covariance matrix $\frac{1}{2}\Delta J$ corresponding to the form

$$\frac{1}{2}j(z, z') = \frac{1}{2}\Delta(z, Jz'). \tag{3.17}$$

Let ρ be the Gaussian state with zero mean and covariance matrix α, and ρ_0 is the pure Gaussian state with zero mean and covariance matrix $\frac{1}{2}\Delta J$. They are both gauge invariant relative to the complex structure J. Then the following decomposition holds

$$\rho = \int W(z)\rho_0 W(z)^* P(d^{2s}z), \qquad (3.18)$$

where P is a J-invariant Gaussian probability measure on Z.

Indeed, the inequality (3.16) implies that the form

$$\alpha(z, z') - \frac{1}{2}j(z, z') \qquad (3.19)$$

is nonnegative definite, hence the characteristic function of the state ρ admit the decomposition

$$\exp\left[-\frac{1}{2}\alpha(z, z)\right] = \varphi(z)\exp\left[-\frac{1}{2}j(z, z')\right], \qquad (3.20)$$

where

$$\varphi(z) = \exp\left[-\frac{1}{2}\left(\alpha(z, z) - \frac{1}{2}j(z, z')\right)\right]$$

is symplectic characteristic function of a Gaussian probability measure P:

$$\varphi(z) = \int \exp(i\Delta(z', z))P(d^{2s}z'). \qquad (3.21)$$

The relation (3.18) then follows from (3.20), by comparing quantum characteristic functions of both sides and taking into account (3.3).

In quantum optics with the preferred complex structure arising from the oscillator Hamiltonian, the gauge-invariant pure state is the vacuum state and the relation (3.18) gives Glauber's P-representation e.g. of a temperature state ρ into coherent states $W(z)\rho_0 W(z)^*$ (see e.g. [27]).

Let Z_A, Z_B be two symplectic spaces and consider a channel Φ: $\mathfrak{T}(\mathcal{H}_A) \to \mathfrak{T}(\mathcal{H}_B)$. The channel is called *Gaussian* if the dual channel satisfies

$$\Phi^*[W_B(z_B)] = W_A(K_B z_B)\exp\left[il(z_B) - \frac{1}{2}\mu(z_B, z_B)\right]. \qquad (3.22)$$

The parameters (K, l, μ) of a quantum Gaussian channel satisfy the condition (see [4])

$$\mu \geq \frac{i}{2}\left[\Delta_B - K^\top \Delta_A K\right]. \qquad (3.23)$$

Any Gaussian channel has the covariance property

$$\Phi[W_A(z)\rho_0 W_A(z)^*] = W_B(K'z)\Phi[\rho_0]W_B(K'z)^* \qquad (3.24)$$

where K' is symplectic transpose.

Assume that in Z_A, Z_B operators of complex structure J_A, J_B are fixed, and let G_A, G_B be the corresponding gauge operators in $\mathcal{H}_A, \mathcal{H}_B$ acting according (3.12). Channel Φ is called *gauge covariant*, if

$$\Phi[e^{i\varphi G_A}\rho e^{-i\varphi G_A}] = e^{i\varphi G_B}\Phi[\rho]e^{-i\varphi G_B} \qquad (3.25)$$

for all input states ρ and real numbers φ. For the Gaussian channel with parameters (K, l, μ) this reduces to

$$l = 0, \quad KJ_B - J_A K = 0, \quad [\Delta_B^{-1}\mu, J_B] = 0.$$

Thus, a natural choice of the complex structure in Z_B is given by any J_B, commuting with the operator $\Delta_B^{-1}\mu$. Existence of such a complex structure is proved similarly to (3.10) with the difference that the matrix μ can be degenerated.

In optimization problems with bounded mean energy a natural complex structure in Z_A is determined by the energy operator $A = R^\top \epsilon R$, namely, J_A is the operator of complex structure in Z_A, commuting with the operator $\epsilon \Delta_A$, so that

$$J_A \epsilon + \epsilon J_A^\top = 0. \qquad (3.26)$$

In the case of usual Hamiltonian of the oscillator system the action of J_A reduces to multiplication by i in the complexification associated with creation-annihilation operators.

3.3. The classical capacity

It is natural to consider the classical capacity of quantum Gaussian channel Φ under the additive input constraint corresponding to the quadratic energy operator $F = R^\top \epsilon R$ with positive-definite matrix ϵ. However finding the classical capacity $C(\Phi, F, E)$ in general depends on the solution of the additivity problem. A natural estimate for $C(\Phi, F, E)$ is given by the quantity $C_\chi(\Phi, F, E)$, defined by the relation (2.14), which coincides with $C(\Phi, F, E)$ in case of additivity. In any case, it gives a lower bound for $C(\Phi, F, E)$.

However even the computation of $C_\chi(\Phi, F, E)$ for Gaussian channels remains in general open problem. At least in [25] it was shown that an optimal ensemble always exists and $C_\chi(\Phi, F, E)$ is given by the relation (2.14) with sup replaced by max.

Consider the following **hypothesis of Gaussian optimal ensembles**: *For a Gaussian channel Φ with the quadratic input energy constraint the maximum in the expression*

$$C_\chi(\Phi, F, E) = \max_{\pi : \mathrm{Tr}\, \bar\rho_\pi F \le E} \chi_\Phi(\pi), \qquad (3.27)$$

where

$$\chi_\Phi(\pi) = H(\Phi(\bar\rho_\pi)) - \int_{\mathfrak{S}(\mathcal{H})} H(\Phi(\rho))\pi(d\rho), \qquad (3.28)$$

is attained by the Gaussian ensemble π, consisting of generalized coherent states $W(z)\rho_0 W(z)^*$, where ρ_0 is a pure Gaussian state, with Gaussian probability distribution $P(d^{2n}z)$.

For such an ensemble the covariance property (3.24) implies $H(\Phi[W(z)\rho_0 W(z)^*]) = H(\Phi[\rho_0])$, and hence

$$\chi_\Phi(\pi) = H(\Phi[\bar\rho_\pi]) - H(\Phi[\rho_0]), \qquad (3.29)$$

which leads us to the **hypothesis of Gaussian minimizer for the output entropy**: *For a Gaussian channel Φ the minimum of the output entropy is attained on a (pure) Gaussian state ρ_0.*

Assume that the channel is gauge-covariant. Then from (2.15), where the expression under the supremum is concave function of ρ by concavity of the entropy and convexity of $\hat{H}_\Phi(\rho)$, it follows that the optimizing state ρ can be chosen gauge-invariant since the averaged gauge-invariant state

$$\bar\rho = \frac{1}{2\pi} \int_0^{2\pi} e^{i\varphi G_A} \rho\, e^{-i\varphi G_A} d\varphi$$

gives at least the same value of the maximized concave function. Moreover, consider the gauge transformations in $\mathcal{H}_A$ and define their action on the generalized ensembles by the formula

$$\pi_\varphi(U) = \pi(\{\rho : e^{i\varphi G_A} \rho\, e^{-i\varphi G_A} \in U\}), \qquad \varphi \in [0, 2\pi],$$

for Borel subsets $U \in \mathfrak{S}(\mathcal{H}_A)$. Generalized ensemble is *gauge invariant*, if $\pi_\varphi \equiv \pi$. By using concavity of the functional $\chi_\Phi(\pi)$ and using averaging of ensembles over φ one shows that maximum in (3.27) is attained on a gauge-invariant ensemble. Again, it follows that the average state of such optimal ensemble is gauge-invariant.

The following proposition relates the two hypotheses for gauge-covariant channels.

Proposition 2. *Let Gaussian channel* Φ *be gauge-covariant with respect to the complex structures* J_A, J_B. *Assume that the minimum of the output entropy is attained on a* G_A-*invariant Gaussian state* ρ_0. *Then the hypothesis of optimal Gaussian ensembles is valid, and the optimal ensemble* π *can be chosen such that the output state* $\bar{\rho}_B = \Phi[\bar{\rho}_\pi]$ *is* G_B-*invariant Gaussian state.*

Proof. Let π be an optimal ensemble. Denote $\bar{\rho}_B = \Phi[\bar{\rho}_\pi]$, then $\operatorname{Tr} \bar{\rho}_\pi F \leq E$ and

$$\begin{aligned}
C_\chi(\Phi, F, E) = \chi_\Phi(\pi) &= H(\bar{\rho}_B) - \hat{H}_\Phi(\bar{\rho}_\pi) \\
&\leq H(\bar{\rho}_B) - \check{H}(\Phi) = H(\bar{\rho}_B) - H(\Phi[\rho_0]), \qquad (3.30)
\end{aligned}$$

by the general inequality $\hat{H}_\Phi(\bar{\rho}_\pi) \geq \check{H}(\Phi)$ and the assumption $H(\Phi[\rho_0]) = \check{H}(\Phi)$.

Consider G_A-invariant state

$$\bar{\rho}_A = \frac{1}{2\pi} \int_0^{2\pi} e^{i\varphi G_A} \bar{\rho}_\pi e^{-i\varphi G_A} d\varphi,$$

then $\operatorname{Tr} \bar{\rho}_A F = \operatorname{Tr} \bar{\rho}_\pi F$. Let now $\tilde{\rho}_A$ be the Gaussian state with the same first and second moments as $\bar{\rho}_A$, then again

$$\operatorname{Tr} \tilde{\rho}_A F = \operatorname{Tr} \bar{\rho}_A F = \operatorname{Tr} \bar{\rho}_\pi F. \qquad (3.31)$$

Moreover, $\tilde{\rho}_A$ is G_A-invariant Gaussian state and $\tilde{\rho}_B = \Phi[\tilde{\rho}_A]$ is G_B-invariant Gaussian state with

$$H(\tilde{\rho}_B) \geq H(\Phi[\bar{\rho}_A]) \geq H(\bar{\rho}_B). \qquad (3.32)$$

Here the first inequality follows from the principle of maximal entropy while the second — from concavity of the entropy. Since ρ_0 is pure G_A-invariant Gaussian state, then one has the decomposition (3.18) for the state $\rho = \tilde{\rho}_A$. Denote by $\tilde{\pi}$ ensemble of generalized coherent states $W(z)\rho_0 W(z)^*$ with Gaussian distribution $P(d^{2s}z)$. Then

$$\begin{aligned}
\tilde{\rho}_B = \Phi[\tilde{\rho}_A] &= \int \Phi[W(z)\rho_0 W(z)^*] P(d^{2s}z) \\
&= \int W(K'z)\Phi[\rho_0]W(K'z)^* P(d^{2s}z)
\end{aligned}$$

by the covariance property (3.24), therefore

$$\hat{H}_\Phi(\tilde{\rho}_A) \leq H(\Phi[\rho_0]) = \check{H}(\Phi).$$

By the relation (3.31), ensemble $\tilde{\pi}$ satisfies the energy constraint. Moreover

$$\chi_\Phi(\tilde{\pi}) = H(\tilde{\rho}_B) - \hat{H}_\Phi(\tilde{\rho}_A) \geq H(\tilde{\rho}_B) - H(\Phi[\rho_0]). \tag{3.33}$$

Bringing together the inequalities (3.30), (3.32), (3.33), we obtain $\chi_\Phi(\tilde{\pi}) \geq \chi_\Phi(\pi) = C_\chi(\Phi, F, E)$, therefore $\tilde{\pi}$ is the optimal ensemble with the required properties. □

Acknowledgments

These notes are based on lectures given by the author at IMS of National University of Singapore during Quantum Information Session of the Program "Mathematical Horizons of Quantum Physics" in August 2008. The author thanks IMS for providing such a nice opportunity.

This work was partially supported by RFBR Grant 09-01-00424.

References

1. R. Alicki and M. Fannes, "Note on multiple additivity of minimal Renyi entropy output of the Werner-Holevo channels," *ArXiv:quant-ph/0407033.*

2. K. M. R. Audenaert and S. L. Braunstein, "On strong superadditivity of the entanglement of formation," *Commun. Math. Phys.*, vol. 246, N3, pp. 443-452, 2004.

3. C. H. Bennett, C. A. Fuchs and J. A. Smolin, "Entanglement-enhanced classical communication on a noisy quantum channel," *Quantum Communication, Computing and Measurement, Proc. QCM96*, Ed. by O. Hirota, A. S. Holevo and C. M. Caves. New York: Plenum, pp. 79-88, 1997.

4. F. Caruso, V. Giovannetti, A. S. Holevo and J. Eisert, "Multi-mode Bosonic Gaussian channels," *New J. Phys.*, vol. 10, 083030, 2008.

5. M.-D. Choi, "Completely positive maps on complex matrices," *Linear Alg. and Its Appl.*, vol. 10, pp. 285-290, 1975.

6. T. Cubitt, A. W. Harrow, D. Leung, A. Montanaro and A. Winter, "Counterexamples to additivity of minimum output p-Renyi entropy for p close to 0," *ArXiv:quant-ph/0712.3628.*

7. N. Datta, A. S. Holevo and Y. M. Suhov, "Additivity for transpose depolarizing channels," *Int. J. Quant. Inform.*, vol. 4, N1, 2006.

8. M. Fukuda, C. King and D. Moser, "Comments on Hastings' additivity counterexample," *ArXiv:quant-ph/0905.3697.*

9. M. Fukuda and M. M. Wolf, "Simplifying additivity problems using direct sum constructions," *J. Math. Phys.*, vol. 48, 072-101, 2007.

10. V. Giovannetti, S. Guha, S. Lloyd, L. Maccone, J. H. Shapiro and H. P. Yuen, *Classical capacity of the lossy bosonic channel: the exact solution*, e-print quant-ph/0308012.

11. V. Giovannetti, S. Lloyd, L. Maccone, J. H. Shapiro and B. J. Yen, "Minimum Réenyi and Wehrl entropies at the output of bosonic channels," *ArXiv:quant-ph/0404037.*

12. M. B. Hastings, "A counterexample to additivity of minimum output entropy," *Nature Physics*, vol. 5, pp. 255-257, 2009; *ArXiv:quant-ph/0809.3972*.

13. M. Hayashi, H. Imai, K. Matsumoto, M.-B. Ruskai and T. Shimono, "Qubit channels which require four inputs to achieve capacity: Implications for additivity conjectures," *ArXiv:quant-ph/0403176*.

14. P. Hayden, "The maximal p-norm multiplicativity conjecture is false," *ArXiv:0707.3291*.

15. P. Hayden, D. Leung, P. W. Shor and A. Winter, "Randomizing quantum states: Constructions and applications," *Communications in Mathematical Physics*, vol. 250, pp. 371-391, 2004.

16. P. Hayden, D. W. Leung and A. Winter, "Aspects of generic entanglement," *Communications in Mathematical Physics*, vol. 265, pp. 95-117, 2006. *arXiv:quant-ph/0407049*.

17. P. Hayden and A. Winter, "Counterexamples to the maximal p-norm multiplicativity conjecture for all $p > 1$," *Comm. Math. Phys.*, vol. 284(1), pp. 263-280, 2008; *arXiv:0807.4753*.

18. T. Hiroshima, "Additivity and multiplicativity properties of some Gaussian channels for Gaussian inputs," *Phys. Rev. A*, vol. 73, 012330, 2006.

19. A. S. Holevo, *Probabilistic and Statistical Aspects of Quantum Theory*, Amsterdam, North Holland, 1982. (2nd Russian edition: Moscow, 2003).

20. A. S. Holevo, "Quantum coding theorems," *Russ. Math. Surveys*, vol. 53, pp. 1295-1331, 1998.

21. A. S. Holevo, "Classical capacities of constrained quantum channel," *Probab. Theory and Appl.*, vol. 48, pp. 359-374, 2003.

22. A. S. Holevo, "On complementary channels and the additivity problem," *ArXiv:quant-ph/0509101*.

23. A. S. Holevo, "The additivity problem in quantum information theory," *Proc. of International Congress of Mathematicians*, Publ. EMS, vol. 3, pp. 999-1018, 2006.

24. A. S. Holevo and M. E. Shirokov, "On Shor's channel extension and constrained channels," *Commun. Math. Phys.*, vol. 249, pp. 417-430, 2004.

25. A. S. Holevo and M. E. Shirokov, "Continuous ensembles and the χ-capacity of infinite-dimensional channels," *ArXiv:quant-ph/0403072*.

26. A. S. Holevo and R. A. Werner, "Evaluating capacities of bosonic Gaussian channels," *Phys. Rev. A*, vol. 63, 032312, 2001.

27. J. R. Klauder and E. C. G. Sudarshan, *Fundamentals of Quantum Optics*, W. A. Benjamin, Inc., NY-Amsterdam, 1968.

28. C. King, "Maximal p-norms of entanglement breaking channels," *ArXiv: quant-ph/0212057*.

29. C. King, "Additivity for a class of unital qubit channels," *ArXiv:quant-ph/0103156*.

30. C. King, "The capacity of the quantum depolarizing channel," *ArXiv:quant-ph/0204172*.

31. C. King, K. Matsumoto, M. Natanson and M. B. Ruskai, "Properties of conjugate channels with applications to additivity and multiplicativity," *ArXiv:quant-ph/0509126*.

32. E. H. Lieb, "Gaussian kernels have only Gaussian maximizers," *Invent. Math.*, vol. 102, pp. 179-208, 1990.

33. E. H. Lieb and W. E. Thirring, "Inequalities for the moments of the eigenvalues of the Schrödinger Hamiltonian and their relation to Sobolev inequalities," *Studies in Math. Phys.*, Ed. by E. H. Lieb, B. Simon and A. Wightman, Princeton University Press, pp. 269-297, 1976.

34. G. G. Magaril-Il'yaev and B. M. Tikhomirov, "Convex analysis: theory and applications," *AMS Transl. of Math. Monographs*, vol. 222, 2003.

35. S. Lloyd and H. Pagels, "Complexity as thermodynamic depth," *Ann. Phys.*, vol. 188, 186-213, 1988.

36. E. Lubkin, "Entropy of n-system from its correlations with k-reservoir," *J. Math. Phys.*, vol. 19, pp. 1028-1031, 1978.

37. K. Matsumoto and A. Yura, "Entanglement cost of antisymmetric states and additivity of capacity of some channels," *ArXiv:quant-ph/0306009*.

38. K. Matsumoto, T. Shimono and A. Winter, "Remarks on additivity of the Holevo channel capacity and of the entanglement of formation," *ArXiv:quant-ph/0206148*.

39. S. Osawa and H. Nagaoka, "Numerical experiments on the capacity of quantum channel with entangled input states," *ArXiv:quant-ph/0007115*.

40. A. A. Pomeransky, "Strong superadditivity of the entanglement of formation follows from its additivity," *Arxiv:quant-ph/0305056*.

41. A. Serafini, J. Eisert and M. M. Wolf, "Multiplicativity of maximal output purities of Gaussian channels under Gaussian inputs," *Phys. Rev. A*, vol. 71, 012320, 2005.

42. M. E. Shirokov, "On entropic quantities related to the classical capacity of infinite dimensional quantum channels," *ArXiv:quant-ph/0411091*.

43. P. W. Shor, "On the number of elements needed in a POVM attaining the accessible information," *ArXiv:quant-ph/0009077*.

44. P. W. Shor, "Additivity of the classical capacity of entanglement-breaking quantum channels," *J. Math. Phys.*, vol. 43, pp. 4334-4340, 2003.

45. P. W. Shor, Equivalence of additivity questions in quantum information theory," *Commun. Math. Phys.*, vol. 246, pp. 453-472, 2004.

46. W. A. Stinespring, "Positive functions on C^*-algebras," *Proc. Amer. Math. Soc.*, vol. 6, pp. 211-311, 1955.

47. A. Wehrl, "General properties of entropy," *Rev. Mod. Phys.*, vol. 50, pp. 221-250, 1978.

48. R. A. Werner and A. S. Holevo, "Counterexample to an additivity conjecture for output purity of quantum channels," *J. Math. Phys.* vol. 43, pp. 4353-4357, 2002.

49. A. Winter, "The maximum output p-norm of quantum channels is not multiplicative for any $p > 2$," *ArXiv:quant-ph/0707.0402*.

50. M. M. Wolf, G. Giedke and J. I. Cirac, "Extremality of Gaussian quantum states," *ArXiv:quant-ph/0509154*.

51. K. Zyczkowski and H. Sommers, "Induced measures in the space of mixed quantum states," *J. Phys. A*, vol. 34, pp. 7111-7125, 2001; *ArXiv:quant-ph/0012101*.

DISSIPATIVELY INDUCED BIPARTITE ENTANGLEMENT

Fabio Benatti

Dipartimento di Fisica Teorica, Università di Trieste
34014 Trieste, Italy
and
Istituto Nazionale di Fisica Nucleare, Sezione di Trieste
34014 Trieste, Italy
E-mail: fabio.benatti@ts.infn.it

Two non-directly interacting qubits with equal oscillation frequencies can be entangled on a short time-scale by a purely dissipative mechanism, that is, by the irreversible reduced dynamics that result from weakly coupling them to a common suitably engineered environment. Depending on the two-point time-correlation functions of the latter, the generated entanglement may also persist asymptotically in time.

1. Introduction

Quantum systems can be considered isolated only if the coupling with the environment which contains them can be neglected; if the coupling is weak but not negligible, one may derive a reduced dynamics for the embedded system alone by eliminating the degrees of freedom of the environment. The resulting time-evolution is irreversible and, under certain assumptions and via certain approximation techniques, may turn out to be even Markovian [1–4]. These systems are known as open quantum systems and evolve according to a so-called quantum dynamical semigroup that incorporates the dissipative and noisy effects due to the environment. Usually, the latter acts as a source of decoherence: in general, the corresponding reduced dynamics irreversibly transforms pure states (one-dimensional projections) into mixtures of pure states (density matrices).

One of the most intriguing aspects of quantum coherence is entanglement, that is the existence of purely quantum mechanical correlations, which has become a central topic in quantum information for its many

applications as a physical resource enabling otherwise impossible information processing protocols [5, 6]. The entanglement content of a state of two
qubits embedded in a same heat bath is generally expected to be spoiled
by decoherence effects; it turns out that this is not the only possibility:
if suitably engineered, the reduced dynamics due to the environment can
entangle an initial separable state of two dynamically independent systems.
Indeed, although not directly interacting between themselves, there can be
an environment mediated generation of quantum correlations between two
systems immersed in it.

This possibility has been demonstrated analytically for two qubits with
a same oscillation frequency [7] and two identical harmonic oscillators [8]
evolving according to a reduced master equation of the typical Lindblad
form [9, 10], obtained via the so-called weak-coupling limit, while for harmonic oscillators in a heat bath of other oscillators it has been derived
numerically from the exact time-evolution [11].

2. Two Qubit Master Equation

The problem we will address in the following regards whether an initial sep-
arable pure state $|\psi\rangle \otimes |\phi\rangle$ of two non-interacting qubits with Hamiltonian

$$H_S = \frac{\omega}{2} \left(\sigma_3^{(1)} + \sigma_3^{(2)} \right) \qquad (2.1)$$

can become entangled when weakly coupled to an environment B, typically,
an infinite heat bath in a thermal equilibrium state ρ_β, via an interaction
of the form

$$H_I = \sum_{a=1}^{2} \sum_{i=1}^{3} \sigma_i^{(a)} \otimes X_i^{(a)}, \qquad (2.2)$$

where $\sigma_{1,2,3}^{(1,2)}$ are the Pauli matrices for the two qubits and $X_i^{(a)}$ are Hermitian bath operators with zero mean, $\text{Tr}(\rho_\beta X_i^{(a)}) = 0$.[a] The total Hamiltonian is thus of the form $H_T = H_S + H_B + \lambda H_I$, with λ a dimensionless
coupling constant and H_B the Hamiltonian relative to the bath degrees of
freedom.

In general, the state of the compound system $S + B$ at time $t > 0$ is a
correlated state $\rho_{SB}(t) = \exp(-itH_T)\rho_{SB} \exp(itH_T)$ from which the state
of the two qubits can be extracted as $\rho(t) = \text{Tr}_B(\rho_{SB}(t))$ by tracing out

---

[a] For sake of simplicity, we shall keep, a finite-system notation, traces, density matrices,
despite the bath having infinitely many degrees of freedom.

the environment degrees of freedom. The time-evolution equation for $\rho(t)$ is complicated, plagued by non-linear and memory effects that make impossible to disentangle a meaningful dynamics for S alone out of the global one for $S + B$. However, if the coupling between S and its environment B is sufficiently weak, $\lambda \ll 1$, one may follow a number of approaches aiming at the derivation of a memoryless, that is Markovian, master equation [1, 2, 4, 9]. One of these is the so-called weak-coupling limit; it consists of a number of approximations: 1) one starts with an initial state $\rho_{SB} = \rho \otimes \rho_\beta$; 2) considers the term of order λ^2 in the Dyson-series expansion of the time-evolution in the interaction representation; 3) goes to the slow time-scale $\tau = t\lambda^2$, typical of the dissipative effects; 4) assumes that on this time-scale the bath time-correlations decay sufficiently rapidly that system and bath are practically disentangled, $\rho_{SB}(\tau/\lambda^2) \simeq \rho(\tau/\lambda^2) \otimes \rho_\beta$ and, finally, 5) performs an ergodic (time) average that eliminates all off-resonant oscillations (rotating-wave approximation). This procedure yields the following Kossakowski-Lindblad master equation:

$$\partial_t \rho(t) = -i\big[H_S + \lambda^2\, H_{12}\,, \rho(t)\big] + \lambda^2\, D[\rho(t)] =: L[\rho(t)]\,. \qquad (2.3)$$

The environment contributes to the generator of the reduced dynamics with an Hamiltonian H_{12} and a purely dissipative term $D[\rho(t)]$; both of them depend on the environment through the two-point time-correlation functions

$$G_{ij}^{(ab)}(t) = \mathrm{Tr}\Big(\rho_\beta X_i^{(a)}\, \mathrm{e}^{itH_B}\, X_j^{(b)} \mathrm{e}^{-itH_B}\Big)\,. \qquad (2.4)$$

The bath-induced Hamiltonian H_{12} consists of three terms; two of them provide a Lamb-shift of the single qubit Hamiltonians, while the third one represents a bath-mediated spin-spin interaction:

$$H_{12}^{int} = \sum_{i,j=1}^{3} h_{ij}^{(12)}\, \sigma_i^{(1)} \sigma_j^{(2)}\,, \qquad (2.5)$$

where the 3×3 matrix $h^{(12)} = [h_{ij}^{(12)}]$ is real, but not necessarily Hermitian, with entries constructed with the Hilbert transforms of the bath two-point correlation functions. Instead, the purely dissipative part reads

$$D[\rho(t)] = \sum_{\substack{a,b=1,2 \\ i,j=1,2,3}} C_{ij}^{(ab)} \left(\sigma_i^{(a)}\, \rho(t)\, \sigma_j^{(b)} - \frac{1}{2}\big\{\sigma_j^{(b)} \sigma_i^{(a)}\,, \rho(t)\big\} \right), \qquad (2.6)$$

where the 3×3 matrices $C^{(ab)} = [C_{ij}^{(ab)}]$ have entries constructed with the Fourier transforms of the two-point correlation functions.

Remark 2.1. The 9×9 Kossakowski matrix $C = \begin{pmatrix} C^{(11)} & C^{(12)} \\ C^{(21)} & C^{(22)} \end{pmatrix}$ turns out to be positive definite; this fact guarantees that the master equation (2.3) generates a semigroup of dynamical maps on the two-qubit density matrices, $\gamma_t = \exp(tL)$, $\rho \mapsto \rho(t) = \gamma_t[\rho]$, for $t \geq 0$, which are completely positive [1–3, 12]. Complete positivity is necessary to guarantee that the reduced dynamics consistently describe a physical process. Indeed, one can always think to couple the given two qubits $(1, 2)$ system with other two qubits $(3, 4)$ (a so-called ancillary system) that are dynamically inert; namely, the time-evolution of the composite system is of the form $\gamma_t \otimes \mathrm{id}$. If the maps γ_t were not completely positive, then there would surely exist an entangled state ρ_{ent} of the composite system $(1, 2) + (3, 4)$ such that $\gamma_t \otimes \mathrm{id}[\rho_{ent}]$ is no longer positive and thus not a state [13].

Certainly, the bath-induced interaction (2.5) contributes to entangle suitable initial separable two-qubit states; what is however interesting is that the same may be true of the purely dissipative part of the generator, $D[\rho(t)]$. This effect hinges upon the non-local character of the action corresponding to the off-diagonal blocks $C^{(12)}$ and $C^{(21)}$ of the Kossakowski matrix and depends on the trade-off between the purely decohering local action of the diagonal blocks.

3. Dissipative Entanglement Generation

We shall first concentrate on the environment induced entanglement of two open qubits at small times and discuss sufficient conditions for an initial separable state $|\psi\rangle \otimes |\phi\rangle$ to become entangled. One knows [5, 6] that a two-qubit state ρ is entangled if and only if under partial transposition T, of the second qubit, say, $\mathrm{id} \otimes T[\rho]$ is no longer positive. In order to make this happen, it is sufficient to consider the expansion of the evolving two-qubit state up to first order in t,

$$\rho(t) \simeq |\psi\rangle\langle\psi| \otimes |\phi\rangle\langle\phi| + t\, L[|\psi\rangle\langle\psi| \otimes |\phi\rangle\langle\phi|]\,,$$

and impose that

$$\langle\Psi|\mathrm{id} \otimes T[L[|\psi\rangle\langle\psi| \otimes |\phi\rangle\langle\phi|]]\Psi\rangle < 0\,, \tag{3.1}$$

for at least one, pure entangled two-qubit state $|\Psi\rangle$ orthogonal to $|\psi\rangle \otimes |\phi\rangle$. In fact, in such a case the first-order expansion of the evolving $\rho(t)$ acquires a negative eigenvalue under partial transposition which means that $\rho(t)$ is

is a projection and vice versa. Therefore, the closer ρ_1 is to a projection, the less entangled is $|\psi\rangle$; this vicinity is usefully measured by the so-called *entaglement of formation* [6], that is by the von Neumann entropy of ρ_1 (or equivalenty of ρ_2 since the reduced density matrices of a bipartite pure state have the same non null eigenvalues):

$$S(\rho_1) = -\frac{1-p}{2}\log\frac{1-p}{2} - \frac{1+p}{2}\log\frac{1+p}{2} \tag{3.5}$$

$$p = \sqrt{1-C^2}, \qquad 0 \le C = 2\,|\alpha\delta - \beta\gamma| \le 1. \tag{3.6}$$

Given the two qubit vector state

$$|\tilde{\psi}\rangle = \sigma_2 \otimes \sigma_2|\psi^*\rangle = -\alpha^*\,|11\rangle + \beta^*\,|10\rangle + \gamma^*\,|10\rangle - \delta^*\,|00\rangle ,$$

where $|\psi^*\rangle$ is the conjugate vector of $|\psi\rangle$ with respect to the basis $|0\rangle, |1\rangle$, it is immediate to check that the 4×4 matrix

$$R = |\psi\rangle\langle\psi|\tilde{\psi}\rangle\langle\tilde{\psi}| . \tag{3.7}$$

has $C^2 = 4\,|\alpha\delta - \beta\gamma|^2$ as positive eigenvalue. The square root of the latter is known as the *concurrence* of the pure state $|\psi\rangle$: when it is maximal, $C = 1$, then $p = 1$ and $S(\rho_1) = \log 2$ is maximal, in which case $|\psi\rangle$ is *maximally entangled*; otherwise, when $C = 0$ is minimal, then $p = 1$, $S(\rho_1) = 0$ and $|\psi\rangle$ is separable.

In conclusion, for two qubit pure states, their entanglement is consistently measured by the concurrence C of which the entanglement is a monotonically increasing function. For two qubit density matrices ρ, the entanglement of formation (3.5) is replaced by

$$E(\rho) = \inf\left\{ \sum_i \lambda_i S(\rho_1^i) : \rho = \sum_i \lambda_i |\psi^i\rangle\langle\psi^i| \right\} , \tag{3.8}$$

that is by the smallest convex combination of the entanglement of formation $S(\rho_1^i) = \mathrm{Tr}_2(|\psi^i\rangle\langle\psi^i|)$ of the pure states $|\psi^i\rangle\langle\psi^i|$ in terms of which ρ can be convexly expanded as $\sum_i \lambda_i |\psi^i\rangle\langle\psi^i|$, $\lambda_i \ge 0$, $\sum_i \lambda_i = 1$. Surprisingly, the variational quantity (3.8) can be expressed as in (3.5) with the concurrence C in (3.6) substituted by

$$C = \max\{0, \lambda_1 - \lambda_2 - \lambda_3 - \lambda_4\} , \tag{3.9}$$

where $\lambda_1 \ge \lambda_2 \ge \lambda_3 \ge \lambda_4$ are the square roots of the (positive) eigenvalues of the 4×4 matrix

$$R = \rho\,\sigma_2^{(1)}\sigma_2^{(2)}\,\rho\,\sigma_2^{(1)}\sigma_2^{(2)} , \tag{3.10}$$

which generalizes the matrix R in (3.7) and similarly has non-negative eigenvalues. In the following, we will deal with density matrices

$$\rho = \begin{pmatrix} a & 0 & 0 & 0 \\ 0 & b & c & 0 \\ 0 & c & d & 0 \\ 0 & 0 & 0 & e \end{pmatrix}, \quad a,b,d,e \geq 0, \quad a+b+d+e=1, \quad bd \geq c^2,$$

written with respect to the orthonormal basis $|\uparrow\uparrow\rangle$, $|\uparrow\downarrow\rangle$, $|\downarrow\uparrow\rangle$, $|\downarrow\downarrow\rangle$. In such a case, the concurrence can readily be computed as

$$C(\rho) = \max\{0, 2(|c| - \sqrt{ae})\} . \tag{3.11}$$

Example 3.4. Consider the purely dissipative reduced dynamics studied in the Appendix. The initial separable pure states $|\uparrow\rangle \otimes |\downarrow\rangle$ and $|\downarrow\rangle \otimes |\uparrow\rangle$ correspond to initial conditions $\rho_{33} = \rho_{44} = 1/2$ and $\rho_{34} = \rho_{43} = \pm 1/2$ with respect to basis used to solve the time-evolution equation. They thus evolve into density matrices (with the notations in the Appendix)

$$\rho(t) = \begin{pmatrix} \rho_{11}(t) & 0 & 0 & 0 \\ 0 & \frac{1+2\rho_{33}(t)\pm e^{-4t}}{4} & \frac{2\rho_{33}(t)-1}{4} & 0 \\ 0 & \frac{2\rho_{33}(t)-1}{4} & \frac{1+2\rho_{33}(t)\mp e^{-4t}}{4} & 0 \\ 0 & 0 & 0 & \rho_{22}(t) \end{pmatrix}.$$

In both cases, the concurrence (3.11) is given by

$$C(t,b) = \left| \rho_{33}(t) - \frac{1}{2} \right| - 2\sqrt{\rho_{11}(t)\rho_{22}(t)} ,$$

when $C(t,b) > 0$, otherwise $C(\rho(t)) = 0$ and the state is no more entangled. In Fig. 1, $C(t,b)$ is plotted as a function of $t \geq 0$ for three values of the parameter b: all three behaviors rapidly saturate showing that the entanglement production stops; further, at a fixed time t, the larger the parameter, the higher the curve.

In the case of the initial separable state $|\uparrow\uparrow\rangle$, its only non-trivial entry with respect to the orthonormal basis used in the Appendix is $\rho_{11} = 1$; therefore it evolves into

$$\rho(t) = \begin{pmatrix} \rho_{11}(t) & 0 & 0 & 0 \\ 0 & \frac{\rho_{33}(t)}{2} & \frac{\rho_{33}(t)}{2} & 0 \\ 0 & \frac{\rho_{33}(t)}{2} & \frac{\rho_{33}(t)}{2} & 0 \\ 0 & 0 & 0 & \rho_{22}(t) \end{pmatrix}.$$

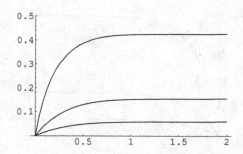

Fig. 1. $C(t, b)$ for $|\uparrow\downarrow\rangle$, $|\downarrow\uparrow\rangle$ and $b = .3$, $b = .5$, $b = .9$.

In this case, $C(t, b) = \rho_{33}(t) - 2\sqrt{\rho_{11}(t)\rho_{22}(t)}$ is never positive (see Fig. 2); therefore, the environment is not able entangle it also at finite times and not only for $t \to 0^+$.

As an initial state, let us now consider $\rho = \frac{1}{2}|\uparrow\downarrow\rangle\langle\uparrow\downarrow| + \frac{1}{2}|\downarrow\uparrow\rangle\langle\downarrow\uparrow|$; it is separable and mixes the first two cases of this example. The dissipative time-evolution sends it into

$$\rho(t) = \frac{1}{2}\begin{pmatrix} \rho_{11}(t) & 0 & 0 & 0 \\ 0 & \frac{1+\rho_{33}(t)}{2} & \frac{\rho_{33}(t)-1}{2} & 0 \\ 0 & \frac{\rho_{33}(t)-1}{2} & \frac{1+\rho_{33}(t)}{2} & 0 \\ 0 & 0 & 0 & \rho_{22}(t) \end{pmatrix},$$

where the entries are as for $|\uparrow\downarrow\rangle$; Fig. 3 plots

$$C(t, b) = \frac{1 - \rho_{33}(t)}{2} - \sqrt{\rho_{11}(t)\rho_{22}(t)}$$

and shows again generation of entanglement at finite times, its increase with increasing values of the parameter b and its saturation when t increases.

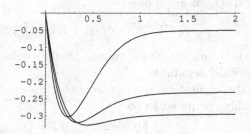

Fig. 2. $C(t, b)$ for $|\uparrow\uparrow\rangle$ and $b = .3$, $b = .5$, $b = .9$.

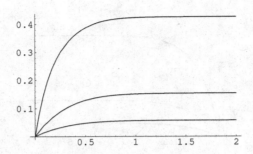

Fig. 3. $C(t,b)$ for $\frac{1}{2}|\uparrow\downarrow\rangle\langle\uparrow\downarrow| + \frac{1}{2}|\downarrow\uparrow\rangle\langle\downarrow\uparrow|$ and $b = .3$, $b = .5$, $b = .9$.

3.1. *Persistence of Entanglement*

The constant value taken up by the concurrence in Figs. 1 and 3 is easily understood by computing the asymptotic state resulting from the explicit time-evolution $t \mapsto \rho(t)$ given in the Appendix: a generic initial two-qubit density matrix ρ goes asymptotically into

$$\rho_\infty = \frac{(1-b)^2}{3+b^2} R |1\rangle\langle 1| + \frac{(1+b)^2}{3+b^2} R |2\rangle\langle 2| + \frac{1-b^2}{3+b^2} R |3\rangle\langle 3| + \rho_{44} |4\rangle\langle 4|$$

$$= \begin{pmatrix} \frac{(1-b)^2 R}{3+b^2} & 0 & 0 & 0 \\ 0 & \frac{(1-b^2)R}{2(3+b^2)} + \frac{\rho_{44}}{2} & \frac{(1-b^2)R}{2(3+b^2)} - \frac{\rho_{44}}{2} & 0 \\ 0 & \frac{(1-b^2)R}{2(3+b^2)} - \frac{\rho_{44}}{2} & \frac{(1-b^2)R}{2(3+b^2)} + \frac{\rho_{44}}{2} & 0 \\ 0 & 0 & 0 & \frac{(1+b)^2 R}{3+b^2} \end{pmatrix}, \qquad (3.12)$$

where $R = \sum_{i=1}^{3} \rho_{ii}$ and ρ_{44} take memory of the initial state (notice that the ρ_{ij} are the coefficients of the expansion of ρ with respect to the or-thonormal basis used in the Appendix, while the matrix is in the standard representation). Then, using that $R = 1 - \rho_{44}$, the concurrence (3.11) of the asymptotic state ρ_∞ is given by

$$C_\infty = \frac{1}{3+b^2} \left(|1 - b^2 - 4\rho_{44}| - 2(1 - b^2)\right), \qquad (3.13)$$

if $C_\infty > 0$ otherwise $C(\rho_\infty) = 0$ and, though initially entangled, the state becomes asymptotically separable.

Using (3.13), the asymptotic concurrences of the states studied in Example 3.4 are readily computed to be

$$C_\infty = \frac{2b^2}{3+b^2} = \begin{cases} 0.058252 & \text{if } b = .3 \\ 0.153846 & \text{if } b = .5 \ . \\ 0.425197 & \text{if } b = .9 \end{cases}$$

for the initial pure separable states $|\uparrow\downarrow\rangle$, $|\downarrow\uparrow\rangle$ and the mixed separable state $\frac{1}{2}|\uparrow\downarrow\rangle\langle\uparrow\downarrow| + \frac{1}{2}|\downarrow\uparrow\rangle\langle\downarrow\uparrow|$, while for $|\uparrow\uparrow\rangle$ it turns out that $C_\infty = -\frac{1-b^2}{3+b^2} \leq 0$.

This asymptotic behavior could have been deduced without solving the master equation (2.3), rather by inspecting the zero eigenvalue subspace of its generator: $D[\rho] = 0$. The available theory [17–19] is seldom constructive; however, in the case of a purely dissipative generator as in (3.4), it can constructively be applied [13, 20].

Notice that the maximally entangled Bell-state $|4\rangle\langle4|$ is left invariant by the dissipative time-evolution γ_t considered in the previous examples. This explains the mechanism of entanglement generation and its asymptotic persistence with respect to the initial mixed state

$$\frac{1}{2}|\uparrow\downarrow\rangle\langle\uparrow\downarrow| + \frac{1}{2}|\downarrow\uparrow\rangle\langle\downarrow\uparrow| = \frac{1}{2}|3\rangle\langle3| + \frac{1}{2}|4\rangle\langle4|.$$

When equally mixed, the maximally entangled Bell states $|3\rangle$ and $|4\rangle$ yield a separable state. However, since the maximal entanglement of $|3\rangle\langle3|$ gets rapidly depleted by γ_t as shown in Fig. 4, while $\gamma_t[|4\rangle\langle4|] = |4\rangle\langle4|$, it is decoherence which makes entanglement emerge.

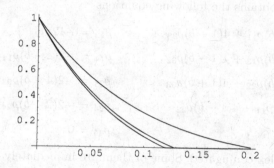

Fig. 4. $C(t, b)$ for $|3\rangle = \frac{|\uparrow\downarrow\rangle + |\downarrow\uparrow\rangle}{\sqrt{2}}$ and $b = .3$, $b = .5$, $b = .9$.

Appendix

In order to solve the master equation (3.4), one first recasts the generator in the form

$$D[\rho(t)] = \sum_{i,j=1}^{3} a_{ij}\left(\Sigma_i\,\rho(t)\,\Sigma_j - \frac{1}{2}\{\Sigma_j\Sigma_i\,,\,\rho(t)\}\right),$$

where $\Sigma_i = \sigma_i^{(1)} + \sigma_i^{(2)}$; and then, after diagonalizing $A = \begin{pmatrix} 1 & ib & 0 \\ -ib & 1 & 0 \\ 0 & 0 & 1 \end{pmatrix}$, one rewrites it as

$$
D[\rho(t)] = 2(1+b)\left(\Sigma_- \rho(t)\Sigma_+ - \frac{1}{2}\{\Sigma_+\Sigma_-\,,\rho(t)\} \right)
$$
$$
+ 2(1-b)\left(\Sigma_+ \rho(t)\Sigma_- - \frac{1}{2}\{\Sigma_-\Sigma_+\,,\rho(t)\} \right)
$$
$$
+ \Sigma_3 \rho(t)\Sigma_3 - \frac{1}{2}\{\Sigma_3^2\,,\rho(t)\}\,,
$$

where $\Sigma_\pm = \sigma_\pm^{(1)} + \sigma_\pm^{(2)}$ and $\sigma_\pm = (\sigma_1 \pm i\sigma_\pm)/2$.

Because of their behavior under the action of $\Sigma_\pm$ and Σ_3, it proves convenient to represent $\rho(t) = \sum_{i,j=1}^4 \rho_{ij}(t)|i\rangle\langle j|$, with respect to the orthonormal basis

$$
|1\rangle = |\uparrow\uparrow\rangle\,, \quad |2\rangle = |\uparrow\uparrow\rangle\,, \quad |3\rangle = \frac{|\uparrow\downarrow\rangle + |\downarrow\uparrow\rangle}{\sqrt{2}}\,, \quad |4\rangle = \frac{|\uparrow\downarrow\rangle - |\downarrow\uparrow\rangle}{\sqrt{2}}\,.
$$

Inserting the expansion of $\rho(t)$ in the right and left hand sides of the master equation, one obtains the following equations

$$\dot{\rho}_{11} = -4(1+b)\rho_{11} + 4(1-b)\rho_{33}\,, \qquad \dot{\rho}_{12} = -12\rho_{12}$$

$$\dot{\rho}_{13} = -2(4+b)\rho_{13} + 4(1-b)\rho_{32}\,, \qquad \dot{\rho}_{14} = -2(2+b)\rho_{14}$$

$$\dot{\rho}_{22} = -4(1-b)\rho_{22} + 4(1+b)\rho_{33}\,, \qquad \dot{\rho}_{23} = -2(4-b)\rho_{23} + 4(1+b)\rho_{31}$$

$$\dot{\rho}_{33} = 4(1+b)\rho_{11} + 4(1-b)\rho_{22} - 8\rho_{33}\,, \qquad \dot{\rho}_{24} = -2(2-b)\rho_{24}$$

$$\dot{\rho}_{34} = -4\rho_{34}\,, \qquad\qquad\qquad \dot{\rho}_{44} = 0\,,$$

plus the complex conjugated. Some of them are immediately integrated,

$$\rho_{12}(t) = \rho_{12}\,e^{-12t}\,, \quad \rho_{14}(t) = \rho_{14}\,e^{-2(2+b)t}\,, \quad \rho_{24}(t) = \rho_{24}\,e^{-2(2-b)t}$$
$$\rho_{34}(t) = \rho_{34}\,e^{-4t}\,, \quad \rho_{44}(t) = \rho_{44}\,.$$

Of the remaining ones, two of them couple the off-diagonal terms ρ_{13} and ρ_{32} yielding

$$
\rho_{13}(t) = \rho_{13}\,F_+(t) + \frac{2(1-b)\rho_{32} - b\rho_{13}}{\sqrt{4-3b^2}}\,F_-(t)
$$

$$
\rho_{32}(t) = \rho_{32}\,F_+(t) + \frac{2(1+b)\rho_{13} + b\rho_{32}}{\sqrt{4-3b^2}}\,F_-(t)\,,
$$

while the other three couple the diagonal entries and obtain

$$
\rho_{11}(t) = \frac{(1-b)^2}{3+b^2} R
$$
$$
+ \sqrt{1-b^2} \frac{(1+b)^2\rho_{11} - 2(1-b)\rho_{22} + (1+b)^2\rho_{33}}{(1+b)(3+b^2)} E_-(t)
$$
$$
+ \frac{2(1+b)\rho_{11} - (1-b)^2(\rho_{22}+\rho_{33})}{3+b^2} E_+(t) \,,
$$
$$
\rho_{22}(t) = \frac{(1+b)^2}{3+b^2} R - \sqrt{1-b^2} \frac{2(1+b)\rho_{11} - (1-b)^2(\rho_{22}+\rho_{33})}{(1-b)(3+b^2)} E_-(t)
$$
$$
- \frac{(1+b)^2\rho_{11} - 2(1+b)\rho_{22} + (1+b)^2\rho_{33}}{3+b^2} E_+(t) \,,
$$
$$
\rho_{33}(t) = \frac{(1-b^2)}{3+b^2} R
$$
$$
+ \sqrt{1-b^2} \frac{(1+b)^3\rho_{11} + (1-b)^3\rho_{22} - 2(1-b^2)\rho_{33}}{(3+b^2)(1-b^2)} E_-(t)
$$
$$
+ \frac{2(1+b^2)\rho_{33} - (1-b^2)(\rho_{11}+\rho_{22})}{3+b^2} E_+(t) \,,
$$

where $R = \rho_{11} + \rho_{22} + \rho_{33} = \rho_{11}(t) + \rho_2(t) + \rho_{33}(t)$ is a constant of the motion and

$$
E_\pm(t) = \frac{e^{-4t(2-\sqrt{1-b^2})} \pm e^{-4t(2+\sqrt{1-b^2})}}{2} \,,
$$
$$
F_\pm(t) = \frac{e^{-2t(4-\sqrt{4-3b^2})} \pm e^{-2t(4+\sqrt{4-3b^2})}}{2}
$$

are quantities which decay asymptotically with $t \to +\infty$. The remaining entries $\rho_{ij}(t)$ follow from complex conjugation.

References

1. H. Spohn, "Kinetic equations from Hamiltonian dynamics: Markovian limits", *Rev. Mod. Phys.*, **52** (1980), 569.
2. R. Alicki and K. Lendi, *Quantum Dynamical Semigroups and Applications*, LNP, **717**, Springer-Verlag, Berlin, 2007.
3. E. B. Davies, "Markovian master equations", *Comm. Math. Phys.*, **39** (1974), 91.
4. H.-P. Breuer and F. Petruccione, *The Theory of Open Quantum Systems*, Oxford University Press, Oxford, 2002.
5. R. Horodecki, P. Horodecki, M. Horodecki and K. Horodecki, "Quantum entanglement", quant-ph/0702225.
6. D. Bruss and G. Leuchs, *Lectures on Quantum Information*, Wiley-VCH, Weinheim, 2007.

7. F. Benatti and R. Floreanini, "Controlling entanglement generation in external quantum fields", *J. Opt. B.*, **7** (2005), S429.
8. F. Benatti and R. Floreanini, "Entangling oscillators through environment noise", *J. Phys. A*, **39** (2006), 2689.
9. V. Gorini, A. Frigerio, M. Verri, A. Kossakowski and E. C. G. Sudarshan, "Properties of quantum Markovian master equations", *Rep. Math. Phys.*, **13** (1978), 149.
10. G. Lindblad, "On the generators of quantum dynamical semigroups", *Comm. Math. Phys.*, **48** (1976), 119.
11. J. P. Paz and A. J. Roncaglia, "Dynamics of the entanglement between two oscillators in the same environment", *Phys. Rev. Lett.*, **100** (2008), 220401.
12. R. Dümcke and H. Spohn, "The proper form of the generator in the weak coupling limit", *Z. Phys.*, **B34** (1979), 419.
13. F. Benatti and R. Floreanini, "Open quantum dynamics: Complete positivity and entanglement", *Int. J. Mod. Phys. B*, **19** (2005), 3063.
14. F. Benatti, R. Floreanini and M. Piani, "Environment induced entanglement in Markovian dissipative dynamics", *Phys. Rev. Lett.*, **91** (2003), 070402.
15. F. Benatti, A. Liguori and A. Nagy, "Environment induced bipartite entanglement", *J. Math. Phys.*, **49** (2008), 042103.
16. W. K. Wootters, "Entanglement of formation of an arbitrary state of two qubits", *Phys. Rev. Lett.*, **80** (1998), 2245.
17. A. Frigerio, "Quantum dynamical semigroups and approach to equilibrium", *Lett. Math. Phys.*, **2** (1977), 79.
18. A. Frigerio, "Stationary states of quantum dynamical semigroups", *Comm. Math. Phys.*, **63** (1977), 269.
19. H. Spohn, "An algebraic condition for the approach to equilibrium of an open N-level system", *Lett. Math. Phys.*, **2** (1977), 33.
20. R. Romano, "Relaxation to equilibrium driven via indirect control in Markovian dynamics", *Phys. Rev. A*, **76** (2207), 042315.

SCATTERING IN NONRELATIVISTIC
QUANTUM FIELD THEORY

Jan Dereziński

Department of Mathematical Methods in Physics
University of Warsaw
Hoza 74, 00-682 Warszawa, Poland
E-mail: jan.derezinski@fuw.edu.pl

The lecture notes are devoted to some topics in scattering theory for certain models inspired by quantum field theory. As a toy example, we describe scattering theory for van Hove Hamiltonians, where all basic objects can be computed exactly. We also sketch the formalism, basic results and some open problems about the so-called Pauli-Fierz Hamiltonians — a class of models describing a small quantum system interacting with a bosonic field, which have an interesting and nontrivial scattering theory.

1. Introduction

The main aim of these lectures is to sketch the formalism and basic results of the scattering theory for certain classes of models inspired by quantum field theory (QFT). We hope that we will convince the readers that this subject has both mathematical elegance and physical relevance.

In our lectures, we mostly consider models that are quite simple. In particular, they always have localized, fast decaying interactions. We are not going to consider relativistic, or even translation invariant models, whose scattering theory is mathematically more dificult, and often problematic.

In Section 2 we describe the standard formalism of scattering theory [24, 28, 32], whose starting point is a pair of operators H and H_0 on a single Hilbert space. Then, in Section 3 we consider scattering theory of Schrödinger operators [28, 10], which, at least in the short range case, is an application of the standard formalism.

Later on we will see that when one wants to study scattering in QFT models, even very simple ones, the standard formalism has to be modified substantially. Therefore, strictly speaking, Sections 2 and 3 do not belong to the main subject of our lectures. Nevertheless, we believe that it is instructive to start with a discussion of these topics, so that the reader can compare them with scattering in QFT.

We use the term "quantum field theory" in a rather broad meaning. Roughly speaking, for the purpose of these lectures, a quantum field theory Hamiltonian is a self-adjoint operator whose definition is based on the formalism of second quantization, involving creation/annihilation operators and Fock spaces. In Section 4 we briefly recall this formalism [5, 7].

In Section 5 we describe in formal terms general principles of scattering in QFT with localized interactions [13, 20, 31]. We explain, in particular, the meaning of renormalization, which in such models is finite and well understood.

In Section 6 we describe scattering theory of a certain exactly solvable class of Hamiltonians — van Hove Hamiltonians [8].

In Section 7 we discuss the so-called representations of the CCR [5, 7]. They arise naturally in the context of scattering theory for bosonic Hamiltonians and allow us to describe some difficult situations typical e.g. for the infra-red problem.

Section 8 is devoted to the scattering theory for a class of Hamiltonians describing a small system interacting with bosonic quantum fields. Following our earlier works, we call them Pauli-Fierz Hamiltonians, although other names can be found in the literature as well. We describe some rigorous results about this subject, as well as some intriguing unsolved problems [10–12, 14, 15].

In our lectures we do not discuss scattering theory for translation invariant QFT models. This subject is more difficult and its rigorous understanding is limited. Let us give a list of what we know rigorously about this subject.

(1) Scattering theory for N-body Schrödinger Hamiltonians is well understood, thanks to the work of Enss, Sigal, Soffer, Graf, the author and others, see [9] and references therein. It can be interpreted as a rather special example of a quantum field theory [6] for a class of Hamiltonians preserving the number of particles.

(2) The Haag-Ruelle theory gives a satsfactory framework for scattering theory in a relativistic quantum field theory satisfying the so-called Haag-Kastler or Wightman axioms in the presence of an isolated shell in the energy-momentum spectrum [23].

(3) Formal perturbative scattering theory for (nonrelativistic) translation invariant QFT models is described in [13, 31].

(4) Compton scattering at weak coupling and small energy has been studied in an interesting paper of Fröhlich, Griesemer and Schlein [15].

2. Basic Abstract Scattering Theory

In this section we recall the standard formalism of scattering theory in an abstract setting. This topic is well known, see e.g. [28, 24, 32]. Later on we will use a different formalism, but we believe that it is instructive to start with the standard approach.

2.1. *Møller and scattering operators*

Suppose that we are given two self-adjoint operators H_0 and $H = H_0 + V$. The *Møller (or wave) operators* (if they exist) are defined as

$$S^{\pm} := \text{s-} \lim_{t \to \pm\infty} e^{itH} e^{-itH_0}.$$

They satisfy $S^{\pm} H_0 = H S^{\pm}$ and are isometric.

The *scattering operator* is introduced as

$$S = S^{+*} S^-.$$

It satisfies $H_0 S = S H_0$. If $\text{Ran}\, S^+ = \text{Ran}\, S^-$, then it is unitary.

Let us note in parenthesis that in the old literature one can sometimes find a scattering operator of a different kind

$$\tilde{S} = S^+ S^{-*}, \tag{2.1}$$

which satisfies $\tilde{S} H = H \tilde{S}$. Both scattering operators are closely related:

$$\tilde{S} = S^- S^* S^{-*}.$$

2.2. *Measurement of observables*

In this and the next subsections we try to describe how the scattering operator leads to measurable quantities. We will call them (*abstract*) *scattering cross-sections*. Let us note that scattering cross-sections in the context of Schrödinger operators or QED are discussed in essentially every textbook on quantum mechanics or quantum field theory. In distinction to those presentations, we will try to do it in an abstract setting, disregarding the concrete form of a quantum system. I find it curious that the abstract formalism of scattering cross-sections is quite complicated and involves a nontrivial condition, which we will call the predictiveness, see (2.6). Note also, that one treats differently the initial time (when the state is prepared) and the final time (when an observable is measured).

We will go back to scattering cross-sections in Subsection 3.2, where we consider them in the context of Schrödinger operators,

Let us start with recalling some of the basic principles of quantum mechanics. Let ρ be the *density matrix* representing a state prepared at time t_-. (Recall that a density matrix is a positive operator of trace 1). Let A be a self-adjoint operator representing an observable measured at time t_+. We learn at basic courses of quantum mechanics that the average outcome of the measurement, which we call the *expectation of the measurement*, equals

$$\operatorname{Tr} \, Ae^{-i(t_+ - t_-)H}\rho e^{i(t_+ - t_-)H}.$$

In realistic situations, it is often difficult to determine the initial state ρ. Typically, the only thing that the experimenter uses to prepare the initial state can be mathematically described by a certain commuting family of self-adjoint operators, which we will call *the control observable*.

Suppose that the control observable has continuous spectrum (which often happens in practice). Then there does not exist a density matrix, which commutes with the control observable, In fact, this follows from the fact that density matrices have pure point spectrum. Therefore, in such a case it is impossible to prepare a state which has a sharp value of the control observable.

Let us try to describe this situation with a more formal language. The control observable will be represented by a $*$-homomorphism

$$C_\infty(X) \ni f \mapsto \gamma(f) \in B(\mathcal{H}), \tag{2.2}$$

where X is a locally compact Hausdorff space and $C_\infty(X)$ denotes the commutative C^*-algebra of continuous functions on X vanishing at infinity. (For example, we can think of X as $\mathbb{R}^d$ and $\gamma(f)$ as $f(D)$, where D denotes the momentum.) We can assume that the $*$-homomorphism γ is injective. If not, $\operatorname{Ker}\gamma = \{f \in C_\infty(X) \ : \ f = 0 \text{ on } Y\}$ for some closed $Y \subset X$, and we can replace in an obvious way $C_\infty(X)$ with $C_\infty(X\backslash Y)$.

It is convenient to extend the $*$-homomorphism γ to a normal $*$-homomorphism, denoted by the same symbol

$$\mathcal{L}^\infty(X) \ni f \mapsto \gamma(f) \in B(\mathcal{H}), \tag{2.3}$$

where $\mathcal{L}^\infty(X)$ denotes the commutative W^*-algebra of bounded Borel functions on X.

Let U be a Borel set in X and let 1_U denote the characteristic function of U. Let B be a self-adjoint operator. We define two real numbers

$$\sigma_+(U,\gamma,B) := \sup\{\operatorname{Tr}\rho B \ : \ \rho \text{ is a density matrix, } \rho = \gamma(1_U)\rho\gamma(1_U)\},$$

$$\sigma_-(U,\gamma,B) := \inf\{\operatorname{Tr}\rho B \ : \ \rho \text{ is a density matrix, } \rho = \gamma(1_U)\rho\gamma(1_U)\}.$$

Clearly, $\sigma_+(U,\rho,B)$, respectively $\sigma_-(U,\rho,B)$, is an increasing, respectively decreasing function of the set U.

Let $x \in X$. We set

$$\sigma_+(x,\gamma,B) := \inf\{\sigma_+(U,\gamma,B) \ : \ x \in U \text{ open in } X\}, \tag{2.4}$$

$$\sigma_-(x,\gamma,B) := \sup\{\sigma_-(U,\gamma,B) \ : \ x \in U \text{ open in } X\}. \tag{2.5}$$

We will say that x is *predictive* for (γ, B) if

$$\sigma_-(x,\gamma,B) = \sigma_+(x,\gamma,B), \tag{2.6}$$

and then we set $\sigma(x,\gamma,B)$ equal to (2.6).

For instance, if x is closed in X (which, by the injectivity of γ implies $\gamma(1_{\{x\}}) \neq 0$), then x is predictive for (γ, B) iff the value of

$$(\Psi|B\Psi) \tag{2.7}$$

does not depend on a normalized vector $\Psi \in \operatorname{Ran}\gamma(1_{\{x\}})$, and then $\sigma(x,\gamma,B)$ equals (2.7).

Let us go back to the situation where the experiment is prepared at time t_-. and the observable A is measured at time t_+. We assume that the experimenter tries to prepare the initial state so that the initial value of the observable given by γ equals x. We also assume that $x \in X$ is predictive for $(\gamma, e^{i(t_+ - t_-)H}Ae^{-i(t_+ - t_-)H})$. Then the expectation of the measurement is close to

$$\sigma\left(x, \gamma, e^{i(t_+ - t_-)H}Ae^{-i(t_+ - t_-)H}\right).$$

2.3. *Physical meaning of the scattering operator*

The physical importance of the scattering theory is based on the fact that in practical situations it takes a long time to prepare states and to measure observables. Scattering theory provides a natural way to take this into account.

Suppose that H_0 is an operator, which is "easy to control" by the experimentalist. Let ρ be a density matrix and A a self-adjoint operator. We assume for the moment that the experimentalist is able to prepare the state $e^{-it_- H_0} \rho e^{it_- H_0}$ at time t_-, and to measure the observable $e^{it_+ H_0} A e^{-it_+ H_0}$ at time t_+. Suppose also that the standard Møller operators exist, and hence the scattering operator S is well defined. Then it is easy to see that, for $t_- \to -\infty$, $t_+ \to \infty$, the expectation of the measurement converges to

$$\text{Tr } A S \rho S^*. \tag{2.8}$$

Thus, in principle, we can determine the full information about the operator S, up to a phase factor, from experiments.

One can argue that the experiment described above is rather difficult to perform for arbitrary A and ρ. Let us modify it to make it more realistic.

Assume that the observable A commutes with H_0. Then

$$e^{it_+ H_0} A e^{-it_+ H_0} = A$$

does not depend on the time of measurement t_+, and thus should be easy to measure.

ρ is a trace class operator, hence there exists an orthonormal basis consisting of its eigenvectors. If $[H_0, \rho] = 0$, then H_0 has pure point spectrum. But in typical situations H_0 has continuous spectrum. Therefore, there are no density matrices commuting with H_0. It is therefore natural to apply the formalism described in the previous subsection.

First we need to choose a control observable. A possible choice would be the free Hamiltonian H_0, or in the language of the previous subsection, the $*$-homomorphism

$$C_\infty(\text{sp } H_0) \ni f \mapsto f(H_0) \in B(\mathcal{H}),$$

given by the functional calculus. Physically, it means the only observable that we control when preparing the initial state is the energy.

In practice, the experimentalist, when preparing the initial state, controls other observables as well (e.g. the momentum). Assume that they can be described by a $*$-homomorphism γ defined on a C^*-algebra $C_\infty(X)$, see (2.2). It is natural to assume that $f(H_0)$ for $f \in C_\infty(\mathbb{R})$ belongs to the range of γ (which means that the free Hamiltonian is one of control observables).

Suppose that the experimentalist prepares the state at time t_- with the control observable γ arbitrarily close to $x \in X$. Then he performs the measurement of the observable A at time t_+. It follows from the definitions (2.4) and (2.5) that the expectation of such a measurement lies inside or very close to the interval

$$\left[\sigma_- \left(x, \gamma, e^{i(t_+ - t_-)H} A e^{-i(t_+ - t_-)H} \right), \sigma_+ \left(x, \gamma, e^{i(t_+ - t_-)H} A e^{-i(t_+ - t_-)H} \right) \right].$$

Let us now take the limits $t_- \to -\infty$, $t_+ \to \infty$. Assume that we are allowed to change the order of relevant limits. Then, for any $\epsilon > 0$, there exists T such that, for $t_- \leq -T$, $T \leq t_+$, the expectation of the measurement lies in

$$[\sigma_- (x, \gamma, S^* AS) - \epsilon, \ \sigma_+ (x, \gamma, S^* AS) + \epsilon].$$

In particular, let us assume also that x is predictive for $(\gamma, S^* AS)$. Then, as $t_- \to -\infty$ and $t_+ \to \infty$, the expectation of the experiment becomes close to

$$\sigma(x, \gamma, S^* AS). \tag{2.9}$$

(2.9) can be called the *scattering cross-section at $x \in X$ for the observable A*.

2.4. *Problem with eigenvalues*

As before, H and H_0 is a pair of self-adjoint operators. It is easy to see that if the standard Møller operators exist and $H_0 \Psi = E \Psi$, then $H \Psi = E \Psi$. Thus, on the subspace spanned by eigenvectors of H_0, the Møller and scattering operators are equal to the identity. Because of that, in practice the standard formalism of scattering theory is usually applied to Hamiltonians H_0 without point spectrum.

In models inspired by QFT, typically, both H_0 and H have ground states, and these ground states are different. Thus, standard scattering theory is not applicable. Instead, one can sometimes try other approaches.

2.5. *Alternative kinds of Møller operators*

There are various possible alternative kinds of Møller operators, which can be used instead of standard ones. Let us describe two of them.

The *strong Abelian Møller operators* are defined as

$$S_{\text{Ab}}^{\pm} := \text{s-}\lim_{\epsilon \searrow 0} \epsilon \int_0^{\infty} e^{-\epsilon t} e^{\pm itH} e^{\mp itH_0} dt.$$

They satisfy $S_{\text{Ab}}^{\pm} H_0 \doteq H S_{\text{Ab}}^{\pm}$, but do not have to be isometric. If the standard Møller operator exists, then so do the Abelian Møller operators, and they coincide.

Another type of Møller operators that can be found in the literature are *adiabatic Møller operators*. To define them we first introduce the *dynamics with an adiabatically switched on interaction*

$$U_\epsilon(0) = 1, \quad \frac{\mathrm{d}}{\mathrm{d}t} U_\epsilon(t) = iU_\epsilon(t)(H_0 + \mathrm{e}^{-\epsilon|t|}V).$$

Then one sets

$$S_{\mathrm{ad}}^\pm := \text{w-}\lim_{\epsilon \searrow 0} \lim_{t \to \pm\infty} U_\epsilon(t)\mathrm{e}^{-itH_0}.$$

One expects that under quite general assumptions $S_{\mathrm{Ab}}^\pm$ coincides with $S_{\mathrm{ad}}^\pm$. In such a case, we will denote them by $S_{\mathrm{ur}}^\pm$. (The subscript "ur" stands for unrenormalized.)

Suppose that the *vacuum amplitude operators* $Z^\pm := S_{\mathrm{ur}}^{\pm*} S_{\mathrm{ur}}^\pm$ have trivial kernels. Then we can define the renormalized Møller operators

$$S_{\mathrm{rn}}^\pm := S_{\mathrm{ur}}^\pm (Z^\pm)^{-1/2}.$$

They also satisfy $S_{\mathrm{rn}}^\pm H_0 = H S_{\mathrm{rn}}^\pm$ and are isometric.

If $\operatorname{Ran} S_{\mathrm{rn}}^+ = \operatorname{Ran} S_{\mathrm{rn}}^-$, then the renormalized scattering operator

$$S_{\mathrm{rn}} = S_{\mathrm{rn}}^{+*} S_{\mathrm{rn}}^-$$

is unitary and $H_0 S_{\mathrm{rn}} = S_{\mathrm{rn}} H_0$.

2.6. Dyson series for Møller and scattering operators

Set $V(t) = \mathrm{e}^{itH_0} V \mathrm{e}^{-itH_0}$. Expanding in formal power series we obtain

$$S_{\mathrm{Ab}}^+ = \lim_{\epsilon \searrow 0} \sum_{n=0}^\infty \int_{\infty > t_n > \cdots > t_1 > 0} i^n \mathrm{e}^{-\epsilon t_n} V(t_n) \cdots V(t_1) \mathrm{d}t_n \cdots \mathrm{d}t_1,$$

$$S_{\mathrm{ad}}^+ = \lim_{\epsilon \searrow 0} \sum_{n=0}^\infty \int_{\infty > t_n > \cdots > t_1 > 0} i^n \mathrm{e}^{-\epsilon(t_n + \cdots + t_1)} V(t_n) \cdots V(t_1) \mathrm{d}t_n \cdots \mathrm{d}t_1.$$

For $S_{\mathrm{ur}} := S_{\mathrm{ur}}^{+*} S_{\mathrm{ur}}^-$, after performing the $\epsilon \searrow 0$ limit we get

$$S_{\mathrm{ur}} = \sum_{n=0}^\infty \int_{\infty > t_n > \cdots > t_1 > -\infty} i^n V(t_n) \cdots V(t_1) \mathrm{d}t_n \cdots \mathrm{d}t_1.$$

After expanding each term in Feynman diagrams, this formal expansion is the usual starting point for analysis of scattering amplitudes in quantum field theory.

2.7. *Other formalisms of scattering theory*

The formalism of scattering theory that we described in this section started from a pair of operators H_0 and H acting on the same Hilbert space. Note that this formalism does not apply to all situations of physical interest, including many QFT models.

Usually, the main aim of scattering theory is to describe a certain *single* self-adjoint Hamiltonian H acting on a Hilbert space $\mathcal{H}$. We will call H and $\mathcal{H}$ the *physical Hamiltonian* and the *physical Hilbert space* respectively. The "free Hamiltonian", or better to say, the "asymptotic Hamiltonian" is not a priori given. It is even not clear that it should act on the same Hilbert space and that it should be the same for the past and future. In fact, part of our job is to guess the *asymptotic Hilbert spaces* $\mathcal{H}^{\pm\,\mathrm{as}}$ as well as the *asymptotic Hamiltonians* $H^{\pm\,\mathrm{as}}$ together with a construction of the Møller operators $S^{\pm} : \mathcal{H}^{\pm\,\mathrm{as}} \to \mathcal{H}$, which should be isometric (preferably unitary), and intertwine the asymptotic and physical Hamiltonians, i.e. $HS^{\pm} = S^{\pm}H^{\pm\,\mathrm{as}}$. I do not know a single formalism that gives a universal recipe how to do this. For various situations one often needs to find it separately. An example of such a formalism is given in Section 8 where we describe scattering theory for Pauli-Fierz Hamiltonians.

Let us mention that a common way to define Møller operators is to introduce appropriate identification operators $J^{\pm} : \mathcal{H}^{\pm\,\mathrm{as}} \to \mathcal{H}$ such that

$$S^{\pm} := \text{s-}\lim_{t\to\infty} e^{itH} J^{\pm} e^{-itH^{\pm\,\mathrm{as}}}. \tag{2.10}$$

Note that the usual scattering operator $S = S^{+*}S^{-}$ maps $\mathcal{H}^{-\,\mathrm{as}}$ into $\mathcal{H}^{+\,\mathrm{as}}$. The alternative scattering operator $\tilde{S} = S^{+}S^{-*}$, introduced in (2.1), acts on the physical space $\mathcal{H}$.

Let us mention some interesting set-ups of scattering theory, which we will not discuss in these notes:

(1) Many-body Schrödinger operators, see e.g. [10].
(2) Local relativistic QFT, the Haag-Ruelle theory, see e.g. [23].
(3) Obstacle scattering for classical waves.

3. Scattering Theory for 2-Body Schrödinger Operators

In this section we describe basic elements of scattering theory for Schrödinger operators [28, 10]. In the short-range case they follow the rules of the standard formalism, outlined in the previous section. In the long-range case a modification is needed.

3.1. *Short-range case*

Consider the Hilbert space $L^2(\mathbb{R}^d)$ and set

$$H_0 = -\frac{1}{2}\Delta, \quad H = -\frac{1}{2}\Delta + V(x).$$

We say that the potential $V(x)$ is short range if

$$|V(x)| \le C(1 + |x|)^{-1-\mu}, \quad \mu > 0. \tag{3.1}$$

Under this assumption one can show that the standard Møller operators $S^\pm := \text{s-lim}_{t\to\pm\infty} e^{itH} e^{-itH_0}$ exist and their ranges equal the absolute continuous spectral subspace of the operator H, denoted $\text{Ran}\, 1_c(H)$. The last statement is called the asymptotic completeness.

We define as usual the scattering operator S and we introduce the T-operator:

$$S = 1 + iT.$$

3.2. *Physical meaning of scattering cross-sections*

Let ξ be the momentum variable. Let $\hat{\xi} = \xi|\xi|^{-1}$ denote the angular variable. Recall that T commutes with H_0. Therefore, the T-operator has the distributional kernel in the momentum representation:

$$T(\xi_+, \xi_-) = \delta(|\xi_+| - |\xi_-|)T(|\xi_+|, \hat{\xi}_+, \hat{\xi}_-).$$

The scattering cross-section at the energy $\lambda^2/2$, incoming angle $\hat{\xi}_-$ and outgoing angle $\hat{\xi}_+$ is defined as

$$\sigma(\lambda, \hat{\xi}_+, \hat{\xi}_-) := |T(\lambda, \hat{\xi}_+, \hat{\xi}_-)|^2. \tag{3.2}$$

It is commonly accepted that the scattering cross-sections are physically the most relevant quantities that are contained in the scattering operator. Let us try to explain their physical meaning, following the idea sketched in Subsection 2.3.

The rough idea of the scattering cross-section is as follows. Suppose that we prepare a state concentrated around the momentum ξ_- and measure the probability of finding the particle of momentum around ξ_+. Assume that the energies are the same: $|\xi_-|^2/2 = |\xi_+|^2/2$. Then the probability of the measurement is proportional to $\sigma(|\xi_+|, \hat{\xi}_+, \hat{\xi}_-)$, at least if the scattering amplitude is well behaved (sufficiently continuous).

Let us make it more precise. Let $D = -i\nabla_x$ denote the momentum operator. Suppose that we want to measure the observable $a(D)$ at time

t_+. At time t_- prepare the state $e^{-it_-H_0}\rho e^{it_-H_0}$, where for simplicity we assume that the density matrix factorizes in the energy and momenta:

$$\rho(\xi_-, \xi_-') = \rho_{en}(|\xi_-|, |\xi_-'|)\rho_{an}(\hat{\xi}_-, \hat{\xi}_-').$$

We also assume that $a(D)\rho = 0$ (so that we measure only scattered states). By (2.8), the expectation of the measurement converges to

$$\int \int \int \overline{T(|\xi_+|, \hat{\xi}_+, \hat{\xi}_-)} a(\xi_+) T(|\xi_+|, \hat{\xi}_+, \hat{\xi}_-')$$

$$\times \rho_{en}(|\xi_+|, |\xi_+|)\rho_{an}(\hat{\xi}_-, \hat{\xi}_-')|\xi_+|^{d-1} d\xi_+ d\hat{\xi}_- d\hat{\xi}_-'. \qquad (3.3)$$

Let us make some additional assumptions. Fix the incoming angle $\eta_- \in S^{d-1}$. Let us assume that $\hat{\xi}_- \mapsto T(|\xi_+|, \hat{\xi}_+, \hat{\xi}_-)$ is continuous at $\hat{\xi}_- = \hat{\eta}_-$, uniformly for $\xi_+ \in \operatorname{supp} a$. Then it is easy to see that, for any $\epsilon > 0$, there exists $\delta > 0$ such that if $\rho_{an}(\hat{\xi}_-, \hat{\xi}_-')$ is supported in the set

$$|\hat{\xi}_- - \hat{\eta}_-| \leq \delta, \qquad |\hat{\xi}_-' - \hat{\eta}_-| \leq \delta,$$

then the expectation value of the measurement (3.3) differs from

$$\int a(\xi_+)\sigma(|\xi_+|, \hat{\xi}_+, \hat{\eta}_-)\rho_{en}(|\xi_+|, |\xi_+|)|\xi_+|^{d-1} d\xi_+$$

$$\times \int \rho_{an}(\hat{\xi}_-, \hat{\xi}_-') d\hat{\xi}_- d\hat{\xi}_-'. \qquad (3.4)$$

by at most ϵ.

Note that the operator T enters (3.4) only through the scattering cross-section. Therefore, scattering cross-sections are sufficient to describe experiments with a well collimated incident beam.

3.3. *Long-range case*

Suppose that the potential satisfies $V = V_l + V_s$ where V_s is short-range (satisfies (3.1)) and

$$|\partial_x^\alpha V_l| \leq C_\alpha (1 + |x|)^{-|\alpha|-\mu}, \quad \mu > 0, \quad |\alpha| = 0, 1, \ldots. \qquad (3.5)$$

We then say that the potential is long range.

It includes the physically relevant Coulomb potential $V(x) = z|x|^{-1}$, where z is the charge.

One can show that for such potentials standard Møller operators in general do not exist. This is one of manifestations of the infra-red problem in quantum physics. Nevertheless, it is possible to compute scattering cross-sections for long range potentials.

There are several methods to do this. The method presented in many quantum mechanics textbooks goes as follows. First one approximates a given long-range potential by a sequence of short-range potentials. E.g. the Coulomb potential is approximated by the Yukawa potentials $V_\mu = ze^{-\mu|x|}|x|^{-1}$. For short-range potentials one can construct Møller and scattering operators, and hence the scattering cross-sections

$$\sigma_\mu(\lambda, \hat{\xi}_1, \hat{\xi}_2)$$

are well defined. Then one shows that there exists

$$\lim_{\mu \searrow 0} \sigma_\mu(\lambda, \hat{\xi}_1, \hat{\xi}_2),$$

which is interpreted as the scattering cross-section for V.

There exist better approaches to the long-range scattering. Instead of the standard Møller operators, one defines the so-called modified Møller operators for long-range potentials, see e.g. [9]. One way to do it, which works for $\mu > \frac{1}{2}$ in (3.5), is as follows. One introduces the function

$$S(t, \xi) = \frac{t\xi^2}{2} + \int_0^t V_1(s\xi)\mathrm{d}s.$$

Then one can show that there exists

$$S_{\mathrm{lr}}^\pm := \text{s-} \lim_{t \to \pm\infty} e^{itH} e^{-iS(t,D)}. \tag{3.6}$$

(3.6) are called modified Møller operators. They are isometric, intertwine the free and full Hamiltonian, that is $S_{\mathrm{lr}}^\pm H_0 = H S_{\mathrm{lr}}^\pm$. They also satisfy asymptotic completeness, in other words Ran $S_{\mathrm{lr}}^\pm = $ Ran $1_{\mathrm{c}}(H)$.

We introduce the modified scattering operator by setting $S_{\mathrm{lr}} := S_{\mathrm{lr}}^{+*} S_{\mathrm{lr}}^-$ and the T-operator by $S_{\mathrm{lr}} = 1 + iT_{\mathrm{lr}}$. We can write the distributional kernel as

$$T_{\mathrm{lr}}(\xi_+, \xi_-) = \delta(|\xi_+| - |\xi_-|)T_{\mathrm{lr}}(|\xi_+|, \hat{\xi}_+, \hat{\xi}_-).$$

Scattering cross-section are defined as

$$\sigma(\lambda, \hat{\xi}_+, \hat{\xi}_-) := |T_{\mathrm{lr}}(\lambda, \hat{\xi}_+, \hat{\xi}_-)|^2. \tag{3.7}$$

3.4. *Freedom of the choice of modified Møller operators*

The main disadvantage of the formalism described above is the fact that in general there is no canonical choice of $S_{\mathrm{lr}}^\pm$. Nevertheless, this arbitrariness is quite limited. If we have two modified Møller operators $S_{\mathrm{lr},1}^{\pm}$ and $S_{\mathrm{lr},2}^{\pm}$, then there exists a phase function $\psi^{\pm}$ such that

$$S_{\mathrm{lr},1}^{\pm} = S_{\mathrm{lr},2}^{\pm} e^{i\psi^{\pm}(D)},$$

where recall that $D = -i\nabla_x$. This arbitrariness disappears in scattering cross-sections, which are canonically defined.

There is, however, another construction, which is unique and canonical. For long-range potentials, there exists self-adjoint operators $D^{\pm}$ such that, for any $g \in C_{\mathrm{c}}(\mathbb{R}^d)$,

$$g(D^{\pm}) = \text{s-}\lim_{t \to \pm\infty} e^{itH} g(D) e^{-itH} 1_{\mathrm{c}}(H).$$

Unlike modified Møller operators, asymptotic momenta are canonically defined. Following [10], one can define canonically the whole class of modified Møller operators as isometric operators $S_{\mathrm{lr}}^{\pm}$ satisfying

$$g(D^{\pm}) = S_{\mathrm{lr}}^{\pm} g(D) S_{\mathrm{lr}}^{\pm *}.$$

4. Second Quantization

In this section we will fix our notation for operators on Fock spaces, which will be the main language in the sequel.

4.1. *Fock spaces*

Let $\mathcal{Z}$ be a Hilbert space. Physically, it will have the meaning of a 1-particle space. On $\otimes^n \mathcal{Z}$ we have the obvious natural action of the permutation group, denoted

$$S_n \ni \sigma \mapsto \Theta(\sigma) \in U(\otimes^n \mathcal{Z}).$$

Let us introduce the orthogonal projections onto symmetric/antisymmetric tensors:

$$\Theta_{\mathrm{s}}^n := \frac{1}{n!} \sum_{\sigma \in S_n} \Theta(\sigma),$$

$$\Theta_{\mathrm{a}}^n := \frac{1}{n!} \sum_{\sigma \in S_n} \operatorname{sgn}\sigma\, \Theta(\sigma).$$

Many concepts are paralel for the symmetric (bosonic) and antisymmetric (fermionic) case. The former will be often denoted by the subscript "s" and the latter by the subscript "a". We will write "s/a" to denote "either s or a".

The *n-particle bosonic/fermionic space* is defined as $\otimes_{s/a}^n \mathcal{Z} := \Theta_{s/a}^n \otimes^n \mathcal{Z}$. The *bosonic/fermionic Fock space* is $\Gamma_{s/a}(\mathcal{Z}) := \oplus_{n=0}^\infty \otimes_{s/a}^n \mathcal{Z}$. The vector $\Omega = 1 \in \otimes_{s/a}^0 \mathcal{Z} = \mathbb{C}$ is called the *vacuum*.

4.2. Creation and annihilation operators

For $f \in \mathcal{Z}$ we define the *creation operator*

$$a^*(f)\Psi := \sqrt{n+1}\,\Theta_{s/a}^{n+1} f \otimes \Psi\,, \quad \Psi \in \otimes_{s/a}^n \mathcal{Z},$$

and the *annihilation operator* $a(f) := (a^*(f))^*$.

Note that traditionally, in most physics textbooks, one uses a somewhat different notation for creation and annihilation operators. One identifies $\mathcal{Z}$ with $L^2(\Xi)$ for some measure space $(\Xi, \mathrm{d}\xi)$. If f equals a function $\Xi \ni \xi \mapsto f(\xi)$, then one writes

$$a^*(f) = \int f(\xi) a^*(\xi) \mathrm{d}\xi\,, \quad a(f) = \int \overline{f}(\xi) a(\xi) \mathrm{d}\xi. \tag{4.1}$$

4.3. Field and Weyl operators

In the bosonic case, for $f \in \mathcal{Z}$ we introduce the *field operators*

$$\phi(f) := \frac{1}{\sqrt{2}}(a^*(f) + a(f))\,,$$

and the *Weyl operators*

$$W(f) := \mathrm{e}^{\mathrm{i}\phi(f)}.$$

For later reference note that

$$(\Omega|W(f)\Omega) = \mathrm{e}^{-\|f\|^2/4}.$$

4.4. Wick quantization

Let $b \in B\big(\otimes_{s/a}^n \mathcal{Z}, \otimes_{s/a}^m \mathcal{Z}\big)$. We would like to define its *Wick quantization*. To this end, it will be convenient to use the traditional notation, which involves an identification of $\mathcal{Z}$ with $L^2(\Xi)$. This identification allows us (at least formally) to represent the operator b by its integral kernel of b, which is a function $b(\xi_1, \ldots, \xi_m, \xi'_n, \ldots, \xi'_1)$ symmetric/antisymmetric in its first and last coordinates. The Wick quantization of the polynomial b will be denoted by

$$B = \int b(\xi_1, \ldots, \xi_m, \xi'_n, \ldots, \xi'_1)$$
$$a^*(\xi_1) \cdots a^*(\xi_m) a(\xi'_n) \cdots a(\xi'_1) \mathrm{d}\xi_1 \cdots \xi_n \mathrm{d}\xi'_m \cdots \mathrm{d}\xi'_1. \tag{4.2}$$

It is the operator whose only nonzero matrix elements are between $k + m$ and $k+n$ particle vectors. For $\Phi \in \otimes_{s/a}^{k+m} \mathcal{Z}$, $\Psi \in \otimes_{s/a}^{k+n} \mathcal{Z}$, the corresponding matrix element equals

$$(\Phi|B\Psi) = \frac{\sqrt{(n+k)!(m+k)!}}{k!}(\Phi|b \otimes 1_{\mathcal{Z}}^{\otimes k}\Psi).$$

Let us remark that the operator (4.2) does not depend on the the choice of the identification of $\mathcal{Z}$ with $L^2(\Xi)$. Moreover, (4.2) is consistent with the usual traditional notation, in particular with (4.1).

4.5. *Second quantization of operators*

For an operator q on $\mathcal{Z}$ we define the operator $\Gamma(q)$ on $\Gamma_{s/a}(\mathcal{Z})$ by

$$\Gamma(q)\Big|_{\otimes_{s/a}^n \mathcal{Z}} := q \otimes \cdots \otimes q\Big|_{\otimes_{s/a}^n \mathcal{Z}}.$$

Similarly, for an operator h we define the operator $\mathrm{d}\Gamma(h)$ by

$$\mathrm{d}\Gamma(h)\Big|_{\otimes_{s/a}^n \mathcal{Z}} := \left(h \otimes 1^{(n-1)\otimes} + \cdots + 1^{(n-1)\otimes} \otimes h\right)\Big|_{\otimes_{s/a}^n \mathcal{Z}}.$$

In the traditional notation, if h is the multiplication operator by $h(\xi)$, then $\mathrm{d}\Gamma(h) = \int h(\xi) a_\xi^* a_\xi \mathrm{d}\xi$.

Note the identity $\Gamma(e^{ith}) = e^{it\mathrm{d}\Gamma(h)}$.

5. Scattering for Hamiltonians of Quantum Field Theory

In this section we describe the basics of scattering theory of QFT Hamiltonians with localized interaction and without the "small system" (see Section 8). Unfortunately, in many cases one has to work with formal power series (see however [11]). Most of the general references on the subject are quite old [13, 20, 31].

5.1. *QFT Hamiltonians*

Typical Hamiltonians of QFT have (at least formally) the form

$$H_\lambda := H_0 + \lambda V, \tag{5.1}$$

where

$$H_0 := \int h(\xi) a^*(\xi) a(\xi) \mathrm{d}\xi, \tag{5.2}$$

$$V := \int \sum_{n,m} v_{n,m}(\xi_1, \ldots, \xi_m, \xi_n', \ldots, \xi_1')$$

$$a^*(\xi_1) \cdots a^*(\xi_m) a(\xi_n') \cdots a(\xi_1') \mathrm{d}\xi_1 \cdots \xi_m \mathrm{d}\xi_1' \cdots \mathrm{d}\xi_n'.$$

The polynomials $v_{n,m}$ should be even in fermionic variables. We will assume that the one-particle energy is $h(\xi) = \sqrt{\xi^2 + m^2}$.

The variable ξ has the interpretation of a 1-particle momentum. Clearly, H_0 is translation invariant. The perturbation V is translation invariant iff it has the form

$$v_{n,m}(\xi_1, \ldots, \xi_m, \xi'_n, \ldots, \xi'_1)$$
$$= \tilde{v}_{n,m}(\xi_1, \ldots, \xi_m, \xi'_n, \ldots, \xi'_1)\delta(\xi_1 + \cdots + \xi_m - \xi'_n - \cdots - \xi'_1).$$

In our notes we will not consider translation invariant interactions. We will always assume that $v_{n,m}(\xi_1, \ldots, \xi_m, \xi'_n, \ldots, \xi'_1)$ are smooth and decay fast in all directions. This simplifying assumption expresses in particular the fact that the interaction is well localized. The scattering theory for such interactions is much easier to study and better understood than that for translation invariant interactions.

We will not worry too much about the self-adjointness of H_λ. If we encounter problems, we will work with formal power series.

Actually, in the case of fermions one can define (5.1) as a self-adjoint operator, since the perturbation is bounded. In the case of bosons, the self-adjointness holds if the perturbation is of degree 1. It is also true for 2nd order perturbation that is sufficiently small. Otherwise it can be proven only under special assumptions (e.g. for spacially cut-off $P(\phi)_2$ interactions [18]).

5.2. *QFT Hamiltonians that do not polarize vacuum*

Suppose that

$$v_{n,0} = v_{0,n} = 0. \tag{5.3}$$

Then Ω is an eigenvector of both H_0 and H, and the standard wave operators exist, at least formally, see e.g. [31].

Unfortunately, physically realistic Hamiltonians often polarize the vacuum, and the standard formalism of scattering theory is inapplicable in these cases.

5.3. *Ground state*

In general, at least formally, H_λ possesses a *ground state* Ω_λ with the *ground state energy* E_λ. They depend on λ in terms of a formal perturbation expansion:

$$\Omega_\lambda = \sum_{n=0}^{\infty} \lambda^n \Omega_n, \qquad E_\lambda = \sum_{n=0}^{\infty} \lambda^n E_n.$$

5.4. *Feynman-Dyson approach*

There exist two basic formalisms for scattering theory of QFT Hamiltonians with localized interaction. The first approach can be traced back to the early works on QED. We will call it the *Feynman-Dyson approach*. It starts with introducing the unrenormalized Møller operators. One can prove their existence, at least as formal power series

$$S_{ur}^{\pm} = \text{s-}\lim_{\epsilon \searrow 0} \epsilon \int_0^{\infty} e^{-\epsilon t} e^{\pm itH} e^{\mp it(H_0 - E)} dt$$

$$= \sum_{n=0}^{\infty} \lambda^n S_{ur,n}^{\pm}.$$

One can also show that the *vacuum amplitude operator* $Z = S_{ur}^{-*} S_{ur}^{-} = S_{ur}^{+*} S_{ur}^{+}$ is proportional to identity and equals $Z = |(\Omega_\lambda | \Omega)|^2$. The *renormalized Møller operators* $S_{rn}^{\pm} := S_{ur}^{\pm} Z^{-1/2}$ are formally unitary and so is the *renormalized scattering operator* $S_{rn} := S_{rn}^{+*} S_{rn}^{-}$.

5.5. *The LSZ formalism*

Instead of the scattering theory based on Møller operators, one can proceed differently. Following Lehman-Symanzik-Zimmermann, one can start by introducing the so-called *asymptotic creation/annihilation operators* defined as the limits

$$a_\lambda^{\pm}(f) := \lim_{t \to \pm\infty} e^{itH} a(e^{-ith} f) e^{-itH},$$

$$a_\lambda^{*\pm}(f) := \lim_{t \to \pm\infty} e^{itH} a^*(e^{-ith} f) e^{-itH}.$$

One can show their existence at least as formal power series. They satisfy the usual *canonical commutation/anticommutation relations (CCR/CAR)*. Moreover, asymptotic annihilation operators kill the perturbed ground state:

$$a_\lambda^{\pm}(f) \Omega_\lambda = 0.$$

The renormalized Møller operators can be defined with help of asymptotic operators

$$S_{rn,\lambda}^{\pm} a^*(f_1) \cdots a^*(f_n) \Omega = a_\lambda^{*\pm}(f_1) \cdots a_\lambda^{*\pm}(f_n) \Omega_\lambda.$$

They are formally unitary and intertwine the CCR/CAR:

$$S_{rn,\lambda}^{\pm} a^*(f) = a_\lambda^{*\pm}(f) S_{rn,\lambda}^{\pm},$$
$$S_{rn,\lambda}^{\pm} a(f) = a_\lambda^{\pm}(f) S_{rn,\lambda}^{\pm}.$$

Note that there is no need for renormalization.

One can construct the alternative renormalized scattering operator $\tilde{S}$ with help of asymptotic operators, even skipping the Møller operators, as the unique (up to a phase factor) unitary operator satisfying

$$\tilde{S}_{rn,\lambda} a_\lambda^{*-}(f) = a_\lambda^{*+}(f) \tilde{S}_{rn,\lambda},$$
$$\tilde{S}_{rn,\lambda} a_\lambda^-(f) = a_\lambda^+(f) \tilde{S}_{rn,\lambda}.$$

6. Scattering Theory of Van Hove Hamiltonians

A *van Hove Hamiltonian* is a self-adjoint operator formally defined as

$$H = \int h(\xi) a^*(\xi) a(\xi) d\xi + \int \overline{z}(\xi) a(\xi) d\xi + \int z(\xi) a^*(\xi) d\xi,$$

where $\xi \mapsto h(\xi) \in [0, \infty[$ describes the energy and $\xi \mapsto z(\xi)$ the interaction. Van Hove Hamiltonians form a very instructive class of operators, whose properties, and in particular the scattering theory, are very well understood [8]. They can also serve as a simple illustration of the infra-red and ultraviolet problem. In our lectures we will not discuss the ultraviolet problem and we will always assume that at high energies the coupling function is sufficiently regular, which is expressed by the condition

$$\int_{h \geq 1} |z(\xi)|^2 d\xi < \infty.$$

Following [8], we will however discuss the infra-red behavior of van Hove Hamiltonians, which is relevant for their scattering theory. One can distinguish 3 cases of the infra-red behavior of the coupling function. In the order of an increasing singularity, we call them A, B and C.

6.1. *Infra-red case A*

We say that the coupling function belongs to Case A if

$$\int_{h < 1} \frac{|z(\xi)|^2}{h(\xi)^2} d\xi < \infty. \tag{6.1}$$

(The integral (6.1) is restricted to ξ with $h(\xi) < 1$.) Van Hove Hamiltonians with the coupling function satisfying this condition are the most regular.

It is easy to see that they are bounded from below self-adjoint operators with the ground state energy

$$E := - \int \frac{|z(\xi)|^2}{h(\xi)} d\xi, \tag{6.2}$$

and the spectrum $[E, \infty[$. Besides, the coherent vector

$$\Psi = \exp\left(- \int \frac{|z(\xi)|^2}{2h(\xi)^2} d\xi\right) \exp\left(\int a^*(\xi) \frac{z(\xi)}{h(\xi)} d\xi\right) \Omega,$$

is its unique ground state.

To see this it is enough to introduce the so-called *dressing operator*

$$U := \exp\left(-a^*\left(\frac{z}{h}\right) + a\left(\frac{z}{h}\right)\right). \tag{6.3}$$

If we set

$$H_0 = \int h(\xi) a_\xi^* a_\xi d\xi,$$

then the operator H is up to a constant unitarily equivalent to H_0:

$$H - E = U H_0 U^*. \tag{6.4}$$

6.2. *Infra-red case B*

Let

$$\int_{h<1} \frac{|z(\xi)|^2}{h(\xi)} d\xi < \infty,$$

$$\int_{h<1} \frac{|z(\xi)|^2}{h(\xi)^2} d\xi = \infty.$$

In this case H can be still defined as a self-adjoint operator and is bounded from below. Equation (6.2) defines a finite number E, which is the infimum of the spectrum of H. However, H has no eigenvalues. This is related to the fact that the dressing operator (6.3) is ill defined, and hence we cannot write (6.4).

6.3. *Infra-red case C*

Let

$$\int_{h<1} |z(\xi)|^2 d\xi < \infty,$$

$$\int_{h<1} \frac{|z(\xi)|^2}{h(\xi)} d\xi = \infty.$$

H can be still defined as a self-adjoint operator. However, H has no eigenvectors and its spectrum covers the whole real line.

For coupling functions satisfying

$$\int_{h<1} |z(\xi)|^2 \mathrm{d}\xi = \infty$$

one cannot define a van Hove Hamiltonian at all.

6.4. Feynman-Dyson scattering theory for van Hove Hamiltonians

Assume that h has an absolutely continuous spectrum (as an operator on $L^2(\Xi)$) and Case A or Case B:

$$\int \frac{|z(\xi)|^2}{h(\xi)} \mathrm{d}\xi < \infty.$$

Then it is easy to show that there exists the strong Abelian Møller operator

$$S_{\mathrm{ur}}^{\pm} := \text{s-}\lim_{\epsilon \searrow 0} \epsilon \int_0^\infty \mathrm{e}^{-\epsilon t} \mathrm{e}^{\mathrm{i}tH} \mathrm{e}^{-\mathrm{i}t(H_0+E)} \mathrm{d}\xi.$$

We have $S_{\mathrm{ur}}^{\pm} = UZ$, where

$$Z = \exp\left(-\int \frac{|z(\xi)|^2}{h^2(\xi)} \mathrm{d}\xi\right).$$

In Case A, the vacuum amplitude constant is nonzero and we can renormalize $S_{\mathrm{ur}}^{\pm}$, obtaining the dressing operator

$$S_{\mathrm{rn}}^{\pm} := S_{\mathrm{ur}}^{\pm} Z^{-1/2} = U.$$

The scattering operator is (unfortunately) trivial:

$$S = S_{\mathrm{rn}}^{+*} S_{\mathrm{rn}}^{-} = 1.$$

In Case B, the vacuum amplitude constant is zero. The Møller operators are not defined. However, if we are willing to introduce and then remove a cut-off, then we can informally conclude that the scattering operator is again equal to identity.

6.5. The LSZ formalism for van Hove Hamiltonians

It is easy to see that in Cases A, B and C, for $f \in \mathrm{Dom}\, h^{-1}$, there exist asymptotic fields:

$$a^{\pm}(f) := \lim_{t \to \pm\infty} \mathrm{e}^{\mathrm{i}tH} a(\mathrm{e}^{-\mathrm{i}th} f) \mathrm{e}^{-\mathrm{i}tH} = a(f) + (f|h^{-1}z),$$

$$a^{*\pm}(f) := \lim_{t \to \pm\infty} \mathrm{e}^{\mathrm{i}tH} a^*(\mathrm{e}^{-\mathrm{i}th} f) \mathrm{e}^{-\mathrm{i}tH} = a^*(f) + (z|h^{-1}f).$$

This allows us to compute the scattering operator $\tilde{S}$ even in Cases B and C. It is trivial — proportional to the identity.

From the point of view of asymptotic fields, the difference between Case A and Cases B and C consists in the type of representations of the CCR: in Case A it is Fock, but in Cases B and C it is not. (Here we use the terminology that we will develop in the next section.)

7. Representations of the CCR

We have seen that the LSZ formalism leads to asymptotic operators satisfying the usual canonical commutation/anticommutation relations (CCR/CAR). These operators can have unusual properties, different from the properties of the usual creation/annihilation operators on a Fock space, as we saw for van Hove Hamiltonians in Cases B and C. Therefore, it is useful to develop a theory of representations of the CCR/CAR in an abstract form. In these lectures we will restrict ourselves to the case of the CCR. We will follow [5, 7].

7.1. *Definition of a representation of the CCR*

Let $\mathcal{Y}$ be a real vector space equipped with an antisymmetric form ω. (Usually we assume that ω is symplectic, i.e. nondegenerate.) Let $U(\mathcal{H})$ denote the set of unitary operators on a Hilbert space $\mathcal{H}$. We say that

$$\mathcal{Y} \ni y \mapsto W^\pi(y) \in U(\mathcal{H})$$

is a *representation of the CCR* over $\mathcal{Y}$ in $\mathcal{H}$ if

$$W^\pi(y_1)W^\pi(y_2) = \mathrm{e}^{-\frac{i}{2}y_1\omega y_2}W^\pi(y_1 + y_2), \qquad y_1, y_2 \in \mathcal{Y}.$$

7.2. *Regular representations of the CCR*

Let $\mathcal{Y} \ni y \mapsto W^\pi(y)$ be a representation of the CCR. Clearly,

$$\mathbb{R} \ni t \mapsto W^\pi(ty) \in U(\mathcal{H})$$

is a 1-parameter group. We say that a representation of the CCR is *regular* if this group is strongly continuous for each $y \in \mathcal{Y}$.

Assume that $y \mapsto W^\pi(y)$ is a regular representation of the CCR. The *field operator* corresponding to $y \in \mathcal{Y}$ is defined as

$$\phi^\pi(y) := -i\frac{\mathrm{d}}{\mathrm{d}t}W^\pi(ty)\Big|_{t=0}.$$

We have the *Heisenberg canonical commutation relations*

$$[\phi^\pi(y_1), \phi^\pi(y_2)] = iy_1 \omega y_2.$$

7.3. Creation/annihilation operators associated with a representation of the CCR

Let $\mathcal{Z}$ be a complex vector space with a scalar product $(\cdot|\cdot)$. It is a symplectic space with the form $\mathrm{Im}(\cdot|\cdot)$. Suppose that

$$\mathcal{Z} \ni f \mapsto W^\pi(f) \in U(\mathcal{H}) \tag{7.1}$$

is a regular representation of the CCR. For $f \in \mathcal{Z}$ we introduce the *creation/ annihilation operators* corresponding to (7.1)

$$a^{\pi*}(f) := \frac{1}{\sqrt{2}}(\phi^\pi(f) + i\phi^\pi(if)), \quad a^\pi(f) := \frac{1}{\sqrt{2}}(\phi^\pi(f) - i\phi^\pi(if)).$$

They satisfy the usual relations

$$[a^\pi(f_1), a^\pi(f_2)] = 0, \quad [a^{\pi*}(f_1), a^{\pi*}(f_2)] = 0,$$
$$[a^\pi(f_1), a^{\pi*}(f_2)] = (f_1|f_2).$$

7.4. The Fock representation

We still consider a complex vector space $\mathcal{Z}$ with a scalar product. Let $\mathcal{Z}^{\mathrm{cpl}}$ denote its completion. Consider the creation/annihilation operators acting on the Fock space $\Gamma_s(\mathcal{Z}^{\mathrm{cpl}})$. Then $\phi(f) := \frac{1}{\sqrt{2}}(a^*(f) + a(f))$ are self-adjoint operators and

$$\mathcal{Z} \ni f \mapsto \exp i\phi(f) \in U\left(\Gamma_s(\mathcal{Z}^{\mathrm{cpl}})\right)$$

is a regular representation of the CCR called the *Fock representation*. The vacuum Ω is characterized by either of the following equivalent equations:

$$a(f)\Omega = 0, \qquad f \in \mathcal{Z};$$
$$(\Omega|e^{i\phi(f)}\Omega) = e^{-\frac{1}{4}(f|f)}, \qquad f \in \mathcal{Z}.$$

7.5. Coherent representations

In this subsection, following [12], we describe an important class of representations of the CCR on a Fock space — coherent representations.

Let g be an antilinear functional on $\mathcal{Z}$ (not necessarily bounded). Then

$$\mathcal{Z} \ni f \mapsto W_g(f) := W(f)e^{i\mathrm{Re}(g|f)} \in U(\Gamma_s(\mathcal{Z}^{\mathrm{cpl}})) \tag{7.2}$$

is a regular representation of the CCR. It will be called *the g-coherent representation*. The corresponding creation/annihilation operators are

$$a_g(f) = a(f) + \frac{1}{\sqrt{2}}(f|g),$$

$$a_g^*(f) = a^*(f) + \frac{1}{\sqrt{2}}(g|f).$$

The vector Ω is characterized by either of the following equations:

$$a_g(f)\Omega = \frac{1}{\sqrt{2}}(f|g)\Omega,$$

$$(\Omega|W_g(f)\Omega) = e^{-\frac{1}{4}(f|f)+i\mathrm{Re}(f|g)}.$$

It is easy to show that the representation $f \mapsto W_g(f)$ is unitarily equivalent to the Fock representation iff g is a bounded functional, equivalently, $g \in \mathcal{Z}^{\mathrm{cpl}}$. More generally, W_{g_1} is equivalent to W_{g_2} iff $g_1 - g_2 \in \mathcal{Z}^{\mathrm{cpl}}$. This gives an obvious equivalence relation on the dual of $\mathcal{Z}$. The equivalence class of g with respect to this relation will be denoted $[g]$.

7.6. *Coherent sectors*

Suppose that

$$\mathcal{Z} \ni f \mapsto W^\pi(f) \in U(\mathcal{H}) \tag{7.3}$$

is a representation of the CCR (e.g. obtained by asymptotic limits, so that $\pi = \pm$). Let g be be an antilinear functional on $\mathcal{Z}$. In this subsection we describe a method that allows us to determine the largest subrepresentation of W^π equivalent to a multiple of the g-coherent representation.

Let $\mathrm{Span}^{\mathrm{cl}}(K)$ denote the closure of the linear span of K. Define

$$\mathcal{K}_g^\pi := \{\Psi \in \mathcal{H} : a^\pi(f)\Psi = \sqrt{2}(g|f)\Psi\} \tag{7.4}$$

$$= \{\Psi \in \mathcal{H} : (\Psi|W^\pi(f)\Psi) = \|\Psi\|^2 e^{-\frac{1}{4}(f|f)+i\mathrm{Re}(f|g)}\},$$

$$\mathcal{H}_{[g]}^\pi := \mathrm{Span}^{\mathrm{cl}}\{a^{\pi*}(f_1)\cdots a^{\pi*}(f_1)\Psi : \Psi \in \mathcal{K}_g^\pi, \ f_i \in \mathcal{Z}\} \tag{7.5}$$

$$= \mathrm{Span}^{\mathrm{cl}}\{W^\pi(f)\Psi : \Psi \in \mathcal{K}_g^\pi, \ f \in \mathcal{Z}\}.$$

$\mathcal{K}_g^\pi$ is called the *space of g-coherent vectors* and $\mathcal{H}_{[g]}^\pi$ is called the $[g]$-*coherent sector* of W^π. In the case $g = 0$, we have a somewhat different terminology: $\mathcal{K}_0^\pi$ is called the *space of Fock vacua* and $\mathcal{H}_{[0]}^\pi$ is called the *Fock sector* of W^π.

We also define an isometric operator $S_g^\pi : \mathcal{K}_g^\pi \otimes \Gamma_s(\mathcal{Z}^{\mathrm{cpl}}) \to \mathcal{H}$ by

$$S_g^\pi \, \Psi \otimes a_g^*(f_1) \cdots a_g^*(f_n)\Omega = a^{\pi*}(f_1) \cdots a^{\pi*}(f_n)\Psi \,,$$

$$S_g^\pi \, \Psi \otimes W_g(f)\Omega = W^\pi(f)\Psi \,. \tag{7.6}$$

(In (7.4), (7.5) and (7.6) we give two alternative equivalent definitions. One of them involves creation/annihilation operators and ther other one involves Weyl operators.)

Theorem 7.1. *The following statements are true:*

(1) $\mathcal{H}_{[g]}^\pi$ *is an invariant subspace for* W^π.
(2) $S_g^\pi : \mathcal{K}_g^\pi \otimes \Gamma_s(\mathcal{Z}^{\mathrm{cpl}}) \to \mathcal{H}_{[g]}^\pi$ *is unitary.*
(3) $S_g^\pi \, 1 \otimes W_g(f) = W^\pi(f) \, S_g^\pi$, $f \in \mathcal{Z}$.
(4) *If* U *is isometric such that* $U \, 1 \otimes W_g(f) = W^\pi(f) \, U$, $f \in \mathcal{Z}$, *then* $\mathrm{Ran}\, U \subset \mathcal{H}_{[g]}^\pi$.

Thus $\mathcal{H}_{[g]}^\pi$ is the biggest subspace of $\mathcal{H}$, on which W^π is unitarily equivalent to W_g.

7.7. Covariant representations

We still consider a representation of the CCR (7.3). Let h be a self-adjoint operator on $\mathcal{Z}^{\mathrm{cpl}}$ and H a self-adjoint operator on $\mathcal{H}$. We say that (W^π, h, H) is a *covariant representation* of the CCR iff

$$\mathrm{e}^{\mathrm{i}tH} W^\pi(f) \mathrm{e}^{-\mathrm{i}tH} = W^\pi(\mathrm{e}^{\mathrm{i}th} f)\,, \quad f \in \mathcal{Z}.$$

The most obvious example of a covariant representation is $(W, h, \mathrm{d}\Gamma(h))$, where W is the Fock representation. This follows from the identity

$$\mathrm{e}^{\mathrm{i}t\mathrm{d}\Gamma(h)} W(f) \mathrm{e}^{-\mathrm{i}t\mathrm{d}\Gamma(h)} = W(\mathrm{e}^{\mathrm{i}th} f).$$

Let us now describe a somewhat more complicated example of a covariant representation. Let $g \in h^{-1}\mathcal{Z}^{\mathrm{cpl}}$. Set $z = \frac{1}{\sqrt{2}}hg$. Introduce the van Hove Hamiltonian

$$\mathrm{d}\Gamma_g(h) := \mathrm{d}\Gamma(h) + a^*(z) + a(z) + (z|h^{-1}z).$$

Let W^g by the g-coherent representation. Then $(W_g, h, \mathrm{d}\Gamma_g(h))$ is a covariant representation of the CCR, that is

$$\mathrm{e}^{\mathrm{i}t\mathrm{d}\Gamma_g(h)} W_g(f) \mathrm{e}^{-\mathrm{i}t\mathrm{d}\Gamma_g(h)} = W_g(\mathrm{e}^{\mathrm{i}th} f). \tag{7.7}$$

Note that (7.7) is obvious for $g \in \mathcal{Z}^{\mathrm{cpl}}$, because then

$$d\Gamma_g(h) = W(ig)d\Gamma(h)W(-ig),$$
$$W_g(f) = W(ig)W(f)W(-ig).$$

7.8. *Coherent sectors of a covariant representation*

The following theorem [12] shows that in some cases subrepresentations of a covariant representation of the CCR are also covariant.

Suppose that $\mathcal{Z} \ni f \mapsto W^\pi(f) \in U(\mathcal{H})$ is a representation of the CCR. We will use the notation $\mathcal{K}_g^\pi$, $\mathcal{H}_{[g]}^\pi$ and S_g^π introduced in (7.4), (7.5) and (7.6).

Theorem 7.2. *Let (W^π, h, H) be covariant. Then the following is true:*

(1) $\mathcal{K}_0^\pi$ *and* $\mathcal{H}_{[0]}^\pi$ *are* e^{itH}*-invariant. Let* $K_0^\pi := H|_{\mathcal{K}_0^\pi}$ *and set*

$$H_0^\pi = K_0^\pi \otimes 1 + 1 \otimes d\Gamma(h).$$

Then $HS_0^\pi = S_0^\pi H_0^\pi$.

(2) *Let* $g \in h^{-1/2}\mathcal{Z}$. *Then* $\mathcal{H}_{[g]}^\pi$ *is* e^{itH}*-invariant. Moreover, for some operator* K_g^π *on* $\mathcal{K}_g^\pi$, *if we set*

$$H_g^\pi := K_g^\pi \otimes 1 + 1 \otimes d\Gamma_g(h),$$

then we have $HS_g^\pi = S_g^\pi H_g^\pi$.

(1) of the above theorem shows that one can always restrict a covariant representation to its Fock sector, obtaining a covariant representation. This covariant representation is very easy — the Hamiltonian restricted to this sector decouples into a sum of non-interacting simple-minded terms.

(2) says that, under some conditions on g, the representation W^π restricted to the $[g]$-coherent sector is still covariant. Moreover, it is unitarily equivalent to

$$\left(1 \otimes W^g, 1 \otimes h, K_g^\pi \otimes 1 + 1 \otimes d\Gamma_g(h)\right).$$

This fact can be used to analyze dynamics that are seemingly difficult, e.g. such as those typical for the infra-red problem [4, 30, 33, 25, 26]. In fact, if $g \notin \mathcal{Z}^{\mathrm{cpl}}$, then the Hamiltonian H restricted to the $[g]$-coherent sector has no eigenvectors, and in spite of that it is under control — its main part is a well understood van Hove Hamiltonian.

8. Pauli-Fierz Hamiltonians

Many physical situations are well described in terms of a "small quantum system" interacting with quantized fields. The small quantum system can be an atom, a molecule, a "quantum dot", etc. One often assumes that it is finite dimensional, or at least that its Hamiltonian has a discrete spectrum. The quantized fields can describe electromagnetic radiation (photons), crystal vibrations (phonons), etc. One often assumes that they are described by a simple free dynamics.

The Hamiltonian of a composite system typically consists of three terms: the Hamiltonian of the small system, the Hamiltonian of the quantum field, and the interaction that couples them.

8.1. *Definition of Pauli-Fierz Hamiltonians*

We will restrict ourselves to the case of bosonic fields and we will assume that the interaction is linear in the fields.

More explicitly, suppose that $\mathcal{K}$ be a Hilbert space with a self-adjoint operator K describing the small system. For instance, we can consider the space $L^2(\mathbb{R}^d)$ with a Schrödinger operator $K = -\Delta + V(x)$. Usually, we will assume that K has discrete eigenvalues, which is the case if $\lim_{|x|\to\infty} V(x) = \infty$.

We assume that the bosons are described by the Fock space $\Gamma_s(\mathcal{Z})$, where, for concreteness, the one-particle space is $\mathcal{Z} = L^2(\mathbb{R}^d)$. As usual, the dispersion relation of the bosons is assumed to be $h(\xi) := \sqrt{\xi^2 + m^2}$, $m \geq 0$. The parameter m will be called "the mass".

The full Hilbert space is $\mathcal{K} \otimes \Gamma_s(\mathcal{Z})$. We fix a coupling function

$$\xi \mapsto v(\xi) \in B(\mathcal{K}).$$

An operator of the form

$$H := H_0 + V, \tag{8.1}$$

where

$$H_0 := K \otimes 1 + 1 \otimes \int h(\xi) a^*(\xi) a(\xi) d\xi, \tag{8.2}$$

$$V := \int v(\xi) \otimes a^*(\xi) d\xi + \text{hc},$$

will be called a *Pauli-Fierz Hamiltonian*. Note in parenthesis that the terminology in this area is not settled and other names are used in this context as well, such as a *generalized spin-boson Hamiltonian*.

8.2. *Spectral properties of Pauli-Fierz Hamiltonians*

Let us start with some results about the spectral properties of Pauli-Fierz Hamiltonians.

Theorem 8.1.

(1) [10] *Assume that* $(K + \mathrm{i})^{-1}$ *is compact and*

$$\int (1 + h(\xi)^{-1})\|v(\xi)\|^2 \mathrm{d}\xi < \infty.$$

Then H *is self-adjoint and bounded from below. If* $E := \inf \mathrm{sp}\, H$, *then*

$$\mathrm{sp}_{\mathrm{ess}}\, H = [E + m, \infty[. \tag{8.3}$$

(2) [16], *see also* [1, 2, 19]. *If in addition*

$$\int (1 + h(\xi)^{-2})\|v(\xi)\|^2 \mathrm{d}\xi < \infty,$$

then H *has a ground state (the infimum of its spectrum is an eigenvalue).*

(1) can be called an HVZ-type theorem for Pauli-Fierz Hamiltonians (after a well known Hunziker-van Winter-Zhislin Theorem about N-body Schrödinger Hamiltonians [29]). It implies that if m is positive, then H necessarily has a ground state. By (2), if the interaction is sufficiently regular in the infrared region, this ground state survives even if $m = 0$.

In typical situations one expects that H has no eigenvalues embedded in its continuous spectrum. This expecation is often confirmed by rigorous results. In fact, for a small non-zero coupling constant and some generic assumptions on the interaction, one can show that the spectrum of $H_\lambda :=$ $H + \lambda V$ in $]E + m, \infty[$ is purely absolutely continuous, e.g. [2, 3].

In particular, if $m = 0$, this means that the only eigenvalue of H_λ is at the bottom of its spectrum. One can often prove that it is nondegenerate.

8.3. *Scattering theory of Pauli-Fierz Hamiltonians*

In the case of Pauli-Fierz Hamiltonians, the formalism of scattering theory based on Abelian Møller operators (which in Section 5 we called the Feynman-Dyson formalism) does not apply. Note that (8.1) is not an operator of the form (5.1), because of the presence of the small system.

It turns out, however, that a certain version of the LSZ formalism works well for Pauli-Fierz Hamiltonians. This formalism will be described below, following its version described by Gérard and the author in [10–12]. (Fröhlich-Griesemer-Schlein use a slightly different setup in [14]).

Theorem 8.2 ([10]). *Suppose that for f from a dense subspace we have*

$$\int_0^\infty \left\| \int e^{ith(\xi)} f(\xi) v(\xi) d\xi + \mathrm{hc} \right\| dt < \infty. \tag{8.4}$$

Define $\mathcal{Z}_1 := \mathrm{Dom}\, h^{-1/2} \subset L^2(\mathbb{R}^d)$. *Then the following holds:*

(1) *For* $f \in \mathcal{Z}_1$, *there exists*

$$W^\pm(f) := \text{s-}\lim_{t\to\pm\infty} e^{itH} \mathbb{1} \otimes W(e^{-ith} f) e^{-itH}; \tag{8.5}$$

(2) $W^\pm(f_1) W^\pm(f_2) = e^{-i\,\mathrm{Im}(f_1|f_2)} W^\pm(f_1 + f_2)$, $f_1, f_2 \in \mathcal{Z}_1$;

(3) $\mathbb{R} \ni t \mapsto W^\pm(tf)$ *is strongly continuous;*

(4) $e^{itH} W^\pm(f) e^{-itH} = W^\pm(e^{ith} f)$;

(5) *If* $H\Psi = E\Psi$, *then* $(\Psi | W^\pm(f)\Psi) = e^{-\|f\|^2/4} \|\Psi\|^2$.

Note that the assumption (8.4) is very weak and it allows for $m = 0$.

Now we can follow the strategy developed in in Section 7. Using *asymptotic Weyl operators* $W^\pm(f)$ we introduce *asymptotic fields*

$$\phi^\pm(f) := \frac{\mathrm{d}}{\mathrm{i}dt} W^\pm(tf)\Big|_{t=0}$$

and *asymptotic creation/annihilation operators*

$$a^{*\pm}(f) := \frac{1}{\sqrt{2}} (\phi(f) + i\phi(if)),$$

$$a^\pm(f) := \frac{1}{\sqrt{2}} (\phi(f) - i\phi(if)).$$

We also define the space of *asymptotic Fock vacua*:

$$\mathcal{K}_0^\pm := \left\{ \Psi \ : \ (\Psi | W^\pm(f)\Psi) = e^{-\|f\|^2/4} \|\Psi\|^2 \right\} \tag{8.6}$$

$$= \left\{ \Psi \ : \ a^\pm(f)\Psi = 0 \right\}. \tag{8.7}$$

(Remember that (8.6) and (8.7) are equal to one another.)

Here is a reformulation of Theorem 8.2, where we use the terminology introduced in Section 7:

Theorem 8.3. *Under the assumptions of Theorem 8.2 the following is true:*

(1) *For* $f \in \mathcal{Z}_1$ *the limit (8.5) exists. Denote it by* $W^\pm(f)$.

(2) $\mathcal{Z}_1 \ni f \mapsto W^\pm(f)$ *are representations of the CCR.*

(3) *These representations are regular.*

(4) $(W^{\pm}, h, H)$ *are covariant.*

(5) $\mathcal{H}_{\mathrm{p}}(H) \subset \mathcal{K}_0^{\pm}$, *where* $\mathcal{H}_{\mathrm{p}}(H)$ *denotes the span of eigenvectors of* H.

8.4. *Asymptotic dynamics*

Let us stress that so far in our scattering theory for Pauli Fierz amiltonians, the starting point was a single Hamiltonian H, and not a pair of Hamiltonians (H, H_0). In fact, *a priori* it is not clear which operator should play the role of the "free Hamiltonian", or better to say, the "asymptotic Hamiltonian". The operator H_0 of (8.2), obtained by dropping the interaction term, is in general not the right choice. In fact, typically, it even has a completely different spectrum than H. In this subsection we will describe how to introduce natural asymptotic Hamiltonians and to construct Møller operators.

First let us introduce the operator

$$K_0^{\pm} := H\big|_{\mathcal{K}_0^{\pm}}.$$

It describes the energies of asymptotic vacua. (Under the assumptions of Theorem 8.4 below we can prove, and under more general conditions we expect, that the spectrum of $K_0^{\pm}$ coincides with the point spectrum of H.)

Define

$$\mathcal{H}_{[0]}^{\pm} := \mathrm{Span}^{\mathrm{cl}} \left\{ W^{\pm}(f)\Psi \ : \ \Psi \in \mathcal{K}_0^{\pm}, \ f \in \mathcal{Z}_1 \right\}.$$

Clearly, $\mathcal{H}_{[0]}^{\pm}$ is the smallest space containing the asymptotic vacua and invariant wrt asymptotic creation operators. It is the largest space on which the asymptotic representations are Fock.

Define the *asymptotic Fock Hilbert space* $\mathcal{H}_0^{\pm\,\mathrm{as}} := \mathcal{K}_0^{\pm} \otimes \Gamma_{\mathrm{s}}(L^2(\mathbb{R}^d))$ and the *asymptotic Hamiltonian for the Fock sector*

$$H_0^{\pm\,\mathrm{as}} := K_0^{\pm} \otimes 1 + 1 \otimes \int h(\xi) a^*(\xi) a(\xi) d\xi.$$

Note that there exist unitary operators

$$S_0^{\pm} : \mathcal{H}_0^{\pm\,\mathrm{as}} \to \mathcal{H}_{[0]}^{\pm} \subset \mathcal{H},$$

which we will call the *Møller operators for the Fock sector*, such that

$$S_0^{\pm}\Psi \otimes a^*(f_1)\cdots a^*(f_n)\Omega = a^{*\pm}(f_1)\cdots a^{*\pm}(f_n)\Psi, \quad \Psi \in \mathcal{K}_0^{\pm}.$$

The Møller operators intertwine the creation/annihilation operators and the Hamiltonian on the asymptotic space, and those on the physical space:

$$S_0^{\pm} 1 \otimes a^*(f) = a^{*\pm}(f) S_0^{\pm},$$
$$S_0^{\pm} 1 \otimes a(f) = a^{\pm}(f) S_0^{\pm},$$
$$S_0^{\pm} H_0^{\pm \, \text{as}} = H S_0^{\pm}.$$

The *scattering operators for the Fock sector* is defined as

$$S_{00} = S_0^{+*} S_0^{-}.$$

It satisfies $S_{00} H_0^{- \, \text{as}} = H_0^{+ \, \text{as}} S_{00}$. If $\mathcal{H}_{[0]}^{+} = \mathcal{H}_{[0]}^{-}$, then S_{00} is unitary on $\mathcal{H}_0^{+ \, \text{as}} = \mathcal{H}_0^{- \, \text{as}}$.

The operator S_{00} can be used to compute various physically interesting scattering cross-sections.

8.5. *Asymptotic completeness*

Theorem 8.2 is not difficult to prove. The following theorem is deeper, especially its second part.

Theorem 8.4 (Asymptotic completeness for massive Pauli-Fierz Hamiltonians). *Assume that $m > 0$. Then*

(1) [21, 10, 11] $\mathcal{H}_{[0]}^{\pm} = \mathcal{H}$, *in other words, the asymptotic representations of the CCR are Fock.*
(2) [10] $\mathcal{K}_0^{\pm} = \mathcal{H}_p(H)$, *in other words, all the asymptotic vacua are linear combinations of eigenvectors.*

In the proof of Theorem 8.4 an important role is played by the methods developed in the study of N-body scattering theory [9]. It is a rather satisfactory result except for one aspect: it assumes the positivity of the mass, which is not very physical. It would be very interesting to extend it to the case $m = 0$. Here is a possible conjecture [12]:

Conjecture. Asymptotic completeness for massless Pauli-Fierz Hamiltonians. *Assume that $h(\xi) = |\xi|$ and*

$$\int (1 + h(\xi)^{-2}) \|v(\xi)\|^2 d\xi < \infty.$$

Then

(1) $\mathcal{H}_{[0]}^{\pm} = \mathcal{H}$,
(2) $\mathcal{K}_0^{\pm} = \mathcal{H}_p(H)$.

Note that the above conjecture is true if $\dim \mathcal{K} = 1$ (i.e. for van Hove Hamiltonians). It is also true if $v(\xi) = 0$ for $|\xi| < \epsilon$, $\epsilon > 0$, (as remarked in [14]).

8.6. *Relaxation to the ground state*

Common wisdom says that a typical small system interacting with a reservoir at zero temperature will relax to its ground state. For a wide and generic class of Pauli-Fierz Hamiltonians this idea can be rigorously expressed and proven, and is essentially an easy corollary of their asymptotic completeness and spectral properties.

As we remarked before, one can often prove that Pauli-Fierz Hamiltonians have only absolutely continuous spectrum except for a unique ground state Ψ_{gr} [2, 3]. If in addition asymptotic completeness holds [10, 14], then the asymptotic space is $\mathcal{H}_0^{\pm \, as} = \Gamma_s(\mathcal{Z})$.

Introduce the C^*-algebra

$$\mathfrak{A} := B(\mathcal{K}) \otimes \mathrm{CCR}(\mathcal{Z}) \subset B\left(\mathcal{K} \otimes \Gamma(\Gamma_s(\mathcal{Z}))\right),$$

where $\mathrm{CCR}(\mathcal{Z}) = \mathrm{Span}^{cl}\{W(f) \; : \; f \in \mathcal{Z}\}$. The following theorem comes from [22, 14]:

Theorem 8.5 (Relaxation to the ground state). *Assume that H is a Pauli-Fierz Hamiltonian for which asymptotic completeness holds, and there are no eigenvectors except for a unique ground state Ψ_{gr}. Let $A \in \mathfrak{A}$. Then*

$$\mathrm{w\text{-}}\lim_{|t| \to \infty} e^{itH} A e^{-itH} = (\Psi_{gr}|A\Psi_{gr}) \; 1_{\mathcal{H}}.$$

8.7. *Coherent asymptotic representations*

In the massless case asymptotic completeness does not always hold. In particular, the Fock property of asymptotic fields may be not true. To see this it is enough to consider the case of van Hove Hamiltonians; more complicated examples can be found in [12]. Nevertheless, following the formalism of Subsection 7.6 and [12], one can try to look for coherent asymptotic representations. This will allow us to study scattering amplitudes also in the case where the Fock property breaks down.

In fact, assume that g belongs to the dual of $\mathcal{Z}_1$. Then one can define the *subspaces of asymptotic g-coherent vectors*

$$\mathcal{K}_g^{\pm} := \left\{ \Psi \in \mathcal{H} \; : \; (\Psi|W^{\pm}(f)\Psi) = \|\Psi\|^2 e^{-\frac{1}{4}(f|f) + i\mathrm{Re}(f|g)} \right\},$$

the [g]-*coherent sector*

$$\mathcal{H}^{\pm}_{[g]} := \mathrm{Span}^{\mathrm{cl}}\left\{W^{\pm}(f)\Psi \ : \ \Psi \in \mathcal{K}^{\pi}_g, \ f \in \mathcal{Z}\right\},$$

the g-*coherent asymptotic Hilbert space*

$$\mathcal{H}^{\pm\,\mathrm{as}}_g := \mathcal{K}^{\pm}_g \otimes \Gamma_{\mathrm{s}}(\mathcal{Z}^{\mathrm{cpl}}),$$

and the g-*coherent asymptotic Hamiltonians*

$$H^{\pm\,\mathrm{as}}_g := K^{\pm}_g \otimes 1 + 1 \otimes \mathrm{d}\Gamma(h).$$

The *Møller operators for the g-coherent sectors* $S^{\pm}_g : \mathcal{H}^{\pm\,\mathrm{as}}_g \to \mathcal{H}^{\pm}_{[g]} \subset \mathcal{H}$ intertwine creation/annihilation operators and the Hamiltonians:

$$S^{\pm}_g 1 \otimes a^*_g(f) = a^{*\pm}(f)S^{\pm}_g,$$
$$S^{\pm}_g 1 \otimes a_g(f) = a^{\pm}(f)S^{\pm}_g,$$
$$S^{\pm}_g H^{\pm\,\mathrm{as}}_g = H S^{\pm}_g.$$

There exists an alternative time-dependent definition of the Møller operator, which follows the pattern (2.10). Define the *g-coherent identifier* $J^{\pm}_g : \mathcal{H}^{\pm\,\mathrm{as}}_g \to \mathcal{H}$ by

$$J^{\pm}_g \ \Psi \otimes W_g(f)\Omega = 1 \otimes W(f) \ \Psi.$$

Then we can introduce the Møller operators using this identifier:

$$S^{\pm}_g = \text{s-}\lim_{t\to\pm\infty} \mathrm{e}^{\mathrm{i}tH} J^{\pm}_g \mathrm{e}^{-\mathrm{i}tH^{\pm\,\mathrm{as}}_g}.$$

Let g_1, g_2 belong to the dual of $\mathcal{Z}_1$. Then one can define the *scattering operator between the sectors corresponding to* g_1 *and* g_2:

$$S_{g_2,g_1} := S^{+*}_{g_2} S^{-}_{g_1}.$$

This operator can be used to define and compute scattering cross-sections even if asymptotic fields have no Fock vacua.

Acknowledgments

The content of this article is based on a tutorial given at IMS of National University of Singapore in September 2008 during the program "Mathematical Horizons of Quantum Physics". The support of IMS is gratefully acknowledged. This work was also partially supported by the Grant N N201 270135.

Some of the results described in this work were obtained jointly with C. Gérard, to whom I owe my gratitude for a fruitful collaboration. I am also thankful to C. A. Pillet for discussions.

References

1. A. Arai and M. Hirokawa, On the existence and uniqueness of ground states of a generalized spin-boson model, J. Func. Anal. 151 (1997) 455.
2. V. Bach, J. Fröhlich and I. Sigal, Quantum electrodynamics of confined non-relativistic particles, Adv. Math. 137 (1998) 299.
3. V. Bach, J. Fröhlich, I. Sigal and A. Soffer, Positive commutators and spectrum of nonrelativistic QED, Commun. Math. Phys. 207 (1999) 557–587.
4. F. Bloch and A. Nordsieck, Note on the radiation field of the electron, Phys. Rev. 52 (1937) 54–59.
5. O. Bratteli and D. W. Robinson, *Operator Algebras and Quantum Statistical Mechanics*, 2nd edition, Springer, Berlin-Heidelberg-New York, Vol. I 1987, Vol. II 1997.
6. J. Dereziński, Asymptotic completeness in quantum field theory. A class of Galilei covariant models, Rev. Math. Phys. 10 (1998) 191–233.
7. J. Dereziński, Introduction to Representations of Canonical Commutation and Anticommutation Relations, *Large Coulomb Systems*, Lecture Notes in Physics 695, eds. J. Dereziński and H. Siedentop, Springer, 2006.
8. J. Dereziński, Van Hove Hamiltonians — exactly solvable models of the infrared an ultraviolet problem, Ann. Henri Poincaré 4 (2003) 713–738.
9. J. Dereziński and C. Gérard, *Scattering Theory of Classical and Quantum N-Particle Systems*, Texts and Monographs in Physics, Springer, 1997.
10. J. Dereziński and C. Gérard, Asymptotic completeness in quantum field theory. Massive Pauli-Fierz Hamiltonians, Rev. Math. Phys. 11 (1999) 383–450.
11. J. Dereziński and C. Gérard, Spectral and scattering theory of spatially cut-off $P(\varphi)_2$ Hamiltonians, Comm. Math. Phys. 213 (2000) 39–125.
12. J. Dereziński and C. Gérard, Scattering theory of infrared divergent Pauli-Fierz Hamiltonians, Ann. Henri Poincaré 5 (2004) 523–577.
13. K. O. Friedrichs, *Perturbations of Spectra in Hilbert Spaces*, AMS Providence, Rhode Island, 1965.
14. J. Fröhlich, M. Griesemer and B. Schlein, Asymptotic completeness for Rayleigh scattering, Ann. Henri Poincaré 3 (2002) 107–170.
15. J. Fröhlich, M. Griesemer and B. Schlein, Asymptotic completeness for Compton scattering, Comm. Math. Phys. 252 (2004) 415–476.
16. C. Gérard, On the existence of ground states for massless Pauli-Fierz Hamiltonians, Ann. Henri Poincaré 1 (2000) 443.
17. C. Gérard, On the scattering theory of massless Nelson models, Rev. Math. Phys. 14 (2002) 1165–1280.
18. J. Glimm and A. Jaffe, *Quantum Physics. A Functional Integral Point of View*, 2nd edition, Springer, New York, 1987.
19. M. Griesemer, E. H. Lieb and M. Loss, Ground states in non-relativistic quantum electrodynamics, Invent. Math. 145 (2001) 557.
20. K. Hepp, *La Theorie de la Renormalisation*, Springer, Berlin-Heidelberg-New York, 1969.
21. R. Hœegh-Krohn, Boson fields under a general class of cut-off interactions, Comm. Math. Phys. 12 (1969) 216–225.

22. M. Hübner and H. Spohn, Radiative decay: Nonperturbative approaches, Rev. Math. Phys. 7 (1995) 363–387.

23. R. Jost, *The General Theory of Quantized Fields*, AMS, Providence, Rhode Island, 1965.

24. T. Kato, *Perturbation Theory for Linear Operators*, 2nd edition, Springer, Berlin, 1984.

25. T. W. B. Kibble, Coherent soft-photon states and infrared divergences I, J. Math. Phys. 9 (1968) 315; II Mass-shell singularities of Green's functions, Phys. Rev. 173 (1968) 1527–1535; III Asymptotic states and reduction formulas, Phys. Rev. 174 (1968) 1882–1901; IV The scattering, Phys. Rev. 175 (1968) 1624–1640.

26. P. P. Kulish and L. D. Faddeev, Asymptotic condition and infrared divergencies in quantum electrodynamics, Theor. Math. Phys. 4 (1970) 745.

27. W. Pauli and M. Fierz, Theory of the emission of long-wave light quanta, Nuovo Cimento 15 (1938) 167–188.

28. M. Reed and B. Simon, *Methods of Modern Mathematical Physics, III. Scattering Theory*, Academic Press, London, 1978.

29. M. Reed and B. Simon, *Methods of Modern Mathematical Physics, IV. Analysis of Operators*, Academic Press, London, 1978.

30. G. Roepstorff, Coherent photon states and spectral condition, Comm. Math. Phys. 19 (1970) 301–314.

31. A. S. Schwarz, *Mathematical Foundations of Quantum Field Theory* (Russian) Nauka, 1975.

32. D. R. Yafaev, *Mathematical Scattering Theory*, Translations of Mathematical Monographs No. 105, AMS, 1992.

33. D. Yennie, S. Frautschi and H. Suura, The infrared divergence phenomena and high-energy processes, Ann. Phys. 13 (1961) 379–452.

MATHEMATICAL THEORY OF ATOMS AND MOLECULES

Volker Bach

Department of Mathematics FB 08
Mainz University
D-55099 Mainz, Germany
E-mail: vbach@mathematik.uni-mainz.de

In these notes, we present some key aspects of the mathematical description of the quantum mechanics of nonrelativistic matter, i.e., atoms and molecules. Special focus lies on the concepts of infinitesimal perturbations, stability of matter, the Hartree-Fock approximation and its justification, and its generalization to a variation over quasifree states which is termed Bogolubov-Hartree-Fock theory.

1. Relative Boundedness and Stability of the First Kind

According to classical mechanics, as it was known about the early 20th century, the electron in a hydrogen atom must fall into the nucleus, emitting an infinite amount of radiation energy. One of the early thriumphes of quantum mechanics in 1925 — as invented by Schrödinger and Heisenberg — was to provide a consistent model in which this does not happen, but the energy of the hydrogen atom is bounded below by a least number — the ground state energy.

Semiboundedness, i.e., boundedness from below, of the energy of a quantum mechanical system became a fundamental principle which has been established for many models and has served to this very day as a guideline to derive new models.

While the semiboundedness of the hydrogen ground state energy was early established by the explicit solution of the Schrödinger equation, a systematic treatment of general interaction potentials that could describe complex molecules or solids was lacking until about 1950, when Kato introduced the notion of relative boundedness.

Definition 1.1. Let $\mathcal{H}$ be a Hilbert space (always complex and separable).

(i) The pair $(A, \mathcal{D})$ is called a **linear operator** on $\mathcal{H}$

$$: \Longleftrightarrow \quad \mathcal{D} \subseteq \mathcal{H} \text{ is a subspace and } A : \mathcal{D} \to \mathcal{H} \text{ is linear.} \quad (1.1)$$

We denote by $\mathcal{L}(\mathcal{H})$ the set of linear operators on $\mathcal{H}$ and call $\mathcal{D}$ **domain** of A.

(ii) A linear operator $(A, \mathcal{D}) \in \mathcal{L}(\mathcal{H})$ is **densely defined**

$$: \Longleftrightarrow \quad \overline{\mathcal{D}} = \mathcal{H}. \quad (1.2)$$

(iii) A linear operator $(A, \mathcal{D}) \in \mathcal{L}(\mathcal{H})$ is **closed**

$$: \Longleftrightarrow \quad \overline{\mathcal{G}(A, \mathcal{D})} = \mathcal{G}(A, \mathcal{D}), \quad \text{where} \quad (1.3)$$
$$\mathcal{G}(A, \mathcal{D}) := \{(\phi, A\phi) \mid \phi \in \mathcal{D}\} \subseteq \mathcal{H} \oplus \mathcal{H}$$

is the **graph** of $(A, \mathcal{D})$.

(iv) For a densely defined linear operator $(A, \mathcal{D})$, we set

$$\mathcal{D}^* := \{\psi \in \mathcal{H} \mid \exists c < \infty \, \forall \phi \in \mathcal{D} : |\langle \psi | A\phi \rangle| \leq c \cdot \|\phi\|\}. \quad (1.4)$$

For $\psi \in \mathcal{D}^*$ the map $(\phi \mapsto \langle \psi | A\phi \rangle) \in \mathcal{B}(\mathcal{H}, \mathbb{C})$ defines a bounded linear functional on $\mathcal{H}$, since $\mathcal{D} \subseteq \mathcal{H}$ is dense. By the Riesz representation theorem, there exists a unique vector $\theta_\psi \in \mathcal{H}$, such that

$$\forall \phi \in \mathcal{D} : \quad \langle \psi | A\phi \rangle = \langle \theta_\psi | \phi \rangle. \quad (1.5)$$

We define **the adjoint** $(A^*, \mathcal{D}^*) \in \mathcal{L}(\mathcal{H})$ of $(A, \mathcal{D})$ by

$$\forall \psi \in \mathcal{D}^* : \quad A^* \psi := \theta_\psi. \quad (1.6)$$

(Linearity of $A^* : \mathcal{D}^* \to \mathcal{H}$ is obvious.)

(v) A densely defined linear operator $(A, \mathcal{D})$ is **symmetric**

$$: \Longleftrightarrow \quad \mathcal{D}^* \supseteq \mathcal{D} \quad \text{and} \quad \forall \varphi \in \mathcal{D} : A^* \varphi = A\varphi. \quad (1.7)$$

In this case, $\langle \varphi | A\varphi \rangle \in \mathbb{R}$, for all $\varphi \in \mathcal{D}$.

(vi) A densely defined, symmetric linear operator $(A, \mathcal{D})$ is **semibounded** or **bounded below**

$$: \Longleftrightarrow \quad \exists C < \infty : \quad A \geq C, \quad (1.8)$$
$$: \Longleftrightarrow \quad \exists C < \infty \, \forall \varphi \in \mathcal{D}, \|\varphi\| = 1 : \langle \varphi | A\varphi \rangle \geq C. \quad (1.9)$$

(vii) A densely defined linear operator $(A, \mathcal{D})$ is **self-adjoint (s.a.)**

$$: \Longleftrightarrow \quad (A^*, \mathcal{D}^*) = (A, \mathcal{D}). \quad (1.10)$$

Examples and Remarks

- Recall that the continuous linear operators on $\mathcal{H}$ are exactly the bounded linear operators on $\mathcal{H}$

$$\mathcal{B}[\mathcal{H}] := \{A : \mathcal{H} \to \mathcal{H} \mid A \text{ is linear and continuous}\}$$
$$= \{A : \mathcal{H} \to \mathcal{H} \mid A \text{ is linear and bounded}\} \quad (1.11)$$
$$= \{A : \mathcal{H} \to \mathcal{H} \mid A \text{ is linear and } \|A\|_{op} < \infty\},$$

where the operator norm $\|A\|_{op}$ is defined as

$$\|A\|_{op} := \sup_{\psi \in \mathcal{H} \setminus \{0\}} \frac{\|A\psi\|_{\mathcal{H}}}{\|A\psi\|_{\mathcal{H}}} := \sup_{\psi \in \mathcal{H} \setminus \{0\}, \|\psi\|_{\mathcal{H}} = 1} \|A\psi\|_{\mathcal{H}}. \quad (1.12)$$

- In quantum mechanics, it is necessary to allow for unbounded operators (for example, the position and momentum operators x and $-i\nabla_x$). Likewise, closedness of the operators is a minimal regularity assumption we require, too.
- The *Closed Graph Theorem* in functional analysis, however, says that if $(A, \mathcal{D}) \in \mathcal{L}(\mathcal{H})$ is a closed linear operator and $\mathcal{D} = \mathcal{H}$ then $A \in \mathcal{B}[\mathcal{H}]$ is bounded.
- Therefore, the domain $\mathcal{D} \subset \mathcal{H}$ of a closed, but unbounded, linear operator $(A, \mathcal{D}) \in \mathcal{L}(\mathcal{H})$ is necessarily a proper subset $\mathcal{D} \neq \mathcal{H}$ of $\mathcal{H}$.

The importance of the notion of self-adjointness of a linear operator $(A, \mathcal{D})$, as opposed to its mere symmetry, lies in Stone's theorem which asserts that the self-adjointness of A is equivalent to the existence and uniqueness of the solution of the initial value problem $\dot{\psi}_t = -iH\psi_t$, $\psi_0 \in \mathcal{H}$, known as Schrödinger's equation. To formulate Stone's theorem, we first introduce the notion of strongly continuous groups.

Definition 1.2. Let $\mathcal{H}$ be a Hilbert space. A family $\{T(t)\}_{t \in \mathbb{R}} \subseteq \mathcal{B}(\mathcal{H})$ of bounded linear operators on $\mathcal{H}$ is called C_0-**group** : $\Longleftrightarrow$

$$\forall t, s \in \mathbb{R} : \quad T(t)T(s) = T(t + s); \quad (1.13)$$
$$T(0) = 1; \quad (1.14)$$
$$\forall \psi \in \mathcal{H} : (t \mapsto T(t)\psi) \in C(\mathbb{R}; \mathcal{H}). \quad (1.15)$$

Theorem 1.3 (Stone (for Self-Adjoint Operators)). *Let $\mathcal{H}$ be a Hilbert space and $(A, \mathcal{D}) \in \mathcal{L}(\mathcal{H})$ a densely defined, closed linear operator on $\mathcal{H}$. Then $(A, \mathcal{D})$ is self-adjoint iff it generates a C_0-group $\{e^{-itA}\}_{t \in \mathbb{R}}$ of unitary operators on $\mathcal{H}$. More precisely,*

(i) *If $(A, \mathcal{D})$ is self-adjoint then the Schrödinger equation*

$$\forall t \in \mathbb{R}: \quad \dot{\psi}_t = -i\, H\, \psi_t \qquad \psi_0 \in \mathcal{D}, \qquad (1.16)$$

has a unique solution $\psi_{(\cdot)} \in C^1(\mathbb{R}; \mathcal{D})$ such that $\|\psi_t\|_{\mathcal{H}} = \|\psi_0\|_{\mathcal{H}}$, for all $t \in \mathbb{R}$. Moreover, the linear operator $T_\mathcal{D}(t) : \mathcal{D} \to \mathcal{D}$ defined by $T_\mathcal{D}(t)\psi_0 := \psi_t$ extends (by continuity) to a unitary operator $T(t) \in \mathcal{B}[\mathcal{H}]$, and the family $\{T(t)\}_{t \in \mathbb{R}}$ constitutes a C_0-group of unitary operators.

(ii) *If $\{T(t)\}_{t \in \mathbb{R}}$ is a given C_0-group of unitary operators, let $\mathcal{D}_T \subseteq \mathcal{H}$ be the subspace of vectors ψ, for which*

$$-i\, A_T\, \psi := \lim_{t \to 0} \left\{ \frac{T(t)\psi - \psi}{t} \right\} \quad exists, \qquad (1.17)$$

*i.e., $\mathcal{D}_T := \{\psi \in \mathcal{H} \mid -iA_T\psi \in \mathcal{H}\}$. Then $\mathcal{D}_T$ is dense, and $(A_T, \mathcal{D}_T) \in \mathcal{L}(\mathcal{H})$ is a self-adjoint operator. In this case A_T is called **generator** of the C_0-group $\{T(t)\}_{t \in \mathbb{R}}$.*

Examples and Remarks

- Theorem 1.3 emphazises the importance of self-adjointness as a property of linear operators. At this point it is not clear, however, that there are any examples.
- The following example illustrates that practically every multiplication operator by a real function is self-adjoint on its natural domain. Namely, let $(\Omega, \mathfrak{A}, \mu)$ be a sigma-finite measure space and $f : \Omega \to \overline{\mathbb{R}}$ a measurable function which is almost everywhere finite, which means that $\lim_{n \to \infty} \mu(\{\omega \in \Omega : |f(\omega)| \geq n\}) = 0$. We set $\mathcal{H} := L^2(\Omega)$,

$$\mathcal{D} := \{\phi \in \mathcal{H} \mid f\phi \in \mathcal{H}\} \qquad (1.18)$$

and $A : \mathcal{D} \to \mathcal{H}$,

$$\forall \omega \in \Omega: \quad (A\phi)(\omega) := f(\omega) \cdot \phi(\omega). \qquad (1.19)$$

Then $(A, \mathcal{D}) \in \mathcal{L}(\mathcal{H})$ is self-adjoint, and its spectrum is given by

$$\sigma(A) = \mathrm{ess\,Ran}(f) := \cap \{\overline{f(\Omega')} \mid \Omega' \in \mathfrak{A},\ \mu(\Omega \setminus \Omega') = 0\}. \qquad (1.20)$$

- The **spectral theorem** for self-adjoint operators says that, given a Hilbert space $\mathcal{H}$ and a self-adjoint operator $(A, \mathcal{D}) = (A^*, \mathcal{D}^*) \in \mathcal{L}(\mathcal{H})$ on $\mathcal{H}$, there exists a measure space $(\Omega, \mathfrak{A}, \mu)$, a real, measurable, and almost everywhere finite, function $f : \Omega \to \overline{\mathbb{R}}$, and a unitary linear map $U : \mathcal{H} \to L^2(\Omega, d\mu)$, such that the domain $\mathcal{D} = \{\phi \in \mathcal{H} \mid (f \cdot U\phi) \in L^2(\Omega, d\mu)\}$

is naturally preserved under U and that, after conjugation with U, the operator A becomes a multiplication operator,

$$\forall \psi \in L^2(\Omega), \quad f \cdot \psi \in L^2(\Omega), \quad \omega \in \Omega: \ (UAU^*\psi)(\omega) = f(\omega)\psi(\omega). \tag{1.21}$$

The spectral theorem hence asserts that *all* self-adjoint operators are multiplication operators, up to a unitary transformation.

- To be more concrete, let $\mathcal{H} = L^2(\mathbb{R}^3)$ and $[|\cdot|^{-1}\varphi](x) := \frac{\varphi(x)}{|x|}$. Then $(|\cdot|^{-1}, \mathcal{D}_{|\cdot|^{-1}})$ is self-adjoint with $\mathcal{D}_{|\cdot|^{-1}} := \{\varphi \in L^2(\mathbb{R}^3) | |\cdot|^{-1}\varphi \in L^2(\mathbb{R}^3)\}$.
- Similarly, let $\mathcal{H} = L^2(\mathbb{R}^3)$ and consider the Laplacian, $[\Delta\varphi](x) := \sum_{\nu=1}^{3}(\partial_\nu^2\varphi)(x)$. Then $(\Delta, \mathcal{D}_\Delta)$ is self-adjoint with $\mathcal{D}_\Delta := \{\varphi \in L^2(\mathbb{R}^3) | \Delta\varphi \in L^2(\mathbb{R}^3)\}$.

  Note that here, ∂_ν are distributional derivatives. For instance, $\partial_\nu\varphi \in L^2(\mathbb{R}^3)$ means that there exists a constant $c < \infty$ such that $|\langle\varphi|\partial_\nu f\rangle| \leq c\|\partial_\nu f\|_{\mathcal{H}}$, for all smooth $f \in C_0^\infty(\mathbb{R}^3)$ of compact support (for which $\partial_\nu f$ is the usual partial derivative w.r.t. x_ν). Since $C_0^\infty(\mathbb{R}^3)$ is dense in $L^2(\mathbb{R}^3)$, there is a unique $\psi \in L^2(\mathbb{R}^3)$ such that $\langle\varphi|\partial_\nu f\rangle = \langle\psi|f\rangle$, for all $f \in C_0^\infty(\mathbb{R}^3)$, and we set $\partial_\nu\varphi := -\psi$ (the minus sign accounting for a ficticious integration by parts). Of course, the distributional derivative of φ agrees with the usual partial derivative, if φ is a differentiable function.
- The previous two examples establish the self-adjointness of the kinetic energy operator $-\Delta$ and the Coulomb potential $-|x|^{-1}$ separately, but this does *not* imply self-adjointness of their sum $-\Delta - |x|^{-1}$. The latter, however, is achieved by the notion of relative boundedness.

Definition 1.4. Let $\mathcal{H}$ be a Hilbert space and $(A, \mathcal{D}_A)$, $(B, \mathcal{D}_B) \in \mathcal{L}(\mathcal{H})$ two densely defined operators on $\mathcal{H}$.

(i) $(B, \mathcal{D}_B)$ is **bounded relative to** A
  $:\Leftrightarrow \quad \mathcal{D}_B \supseteq \mathcal{D}_A$, and there exist $a, b \in \mathbb{R}^+$, such that
  $$\forall \varphi \in \mathcal{D}_A: \quad \|B\varphi\| \leq a \cdot \|A\varphi\| + b \cdot \|\varphi\|. \tag{1.22}$$
  In this case a is called **relative bound**.

(ii) $(B, \mathcal{D}_B)$ is an **infinitesimal perturbation** of $(A, \mathcal{D}_A)$
  $:\Leftrightarrow \quad \mathcal{D}_B \supseteq \mathcal{D}_A$, and for all $a > 0$ there exists $b \equiv b(a) < \infty$, such that
  $$\forall \varphi \in \mathcal{D}_A: \quad \|B\varphi\| \leq a \cdot \|A\varphi\| + b \cdot \|\varphi\|. \tag{1.23}$$

(iii) A symmetric operator $(A, \mathcal{D}_A)$ is **semibounded** or **bounded below**
  $:\Leftrightarrow \quad \exists M > -\infty \ \forall \varphi \in \mathcal{D}_A: \quad \langle\varphi|A\varphi\rangle \geq M \cdot \|\varphi\|^2. \tag{1.24}$
  In this case M is called **lower bound for** A, and we write $A \geq M$.

Theorem 1.5 (Kato-Rellich). *Let $\mathcal{H}$ be a Hilbert space, $(A, \mathcal{D}_A) \in \mathcal{L}(\mathcal{H})$ self-adjoint, $(B, \mathcal{D}_B) \in \mathcal{L}(\mathcal{H})$ symmetric and bounded relative to A with relative bound $a < 1$. Then*

(i) *$(A + B, \mathcal{D}_A) \in \mathcal{L}(\mathcal{H})$ is self-adjoint,*
(ii) *If $(A, \mathcal{D}_A)$ is furthermore semibounded and $A^{\bullet} \geq M$ then $(A + B, \mathcal{D}_A)$ is semibounded, too, and*

$$A + B \geq M - \max\left\{\frac{b}{1-a}, |M|\right\}. \tag{1.25}$$

Examples and Remarks

- Semiboundedness is not only one key condition to ensure self-adjointness by means of the Kato-Rellich theorem, but also reflects the physical aspect of boundedness of the energy of a physical system from below.
- The following lemma gives a convenient characterization for B to be an infinitesimal perturbation of A.

Lemma 1.6. *Let $\mathcal{H}$ be a Hilbert space, $(A, \mathcal{D}_A)$ a self-adjoint and semibounded operator, $(B, \mathcal{D}_B) \in \mathcal{L}(\mathcal{H})$ with $\mathcal{D}_B \supseteq \mathcal{D}_A$ and $B(A+E)^{-1} \in \mathcal{B}(\mathcal{H})$, for $E < \infty$ sufficiently large, such that*

$$\lim_{E \to \infty} \left\| B(A + E)^{-1} \right\|_{\text{op}} = 0. \tag{1.26}$$

Then $(B, \mathcal{D}_B)$ is an infinitesimal perturbation of $(A, \mathcal{D}_A)$.

Examples and Remarks

- We apply Lemma 1.6 to $\mathcal{H} := L^2(\mathbb{R}^d)$, with $d \leq 3$, and $V \in L^2 + L^\infty(\mathbb{R}^d)$. We claim that

$$\lim_{E \to \infty} \left\| V(-\Delta + E)^{-1} \right\|_{\text{op}} = 0. \tag{1.27}$$

To see this we write $V = V_2 + V_\infty$, with $V_p \in L^p(\mathbb{R}^d)$, and let $f, \psi \in L^2(\mathbb{R}^d)$. Then

$$
\begin{aligned}
\left\| V_2(f * \psi) \right\|_{L^2}^2 &= \int |V_2(z)|^2 \left| \int f(x-z)\psi(x) \, d^d x \right|^2 d^d z \\
&\leq \int |V_2(z)|^2 \left(\int |f(x-z)|^2 d^d x \right) \left(\int |\psi(x)|^2 d^d x \right) d^d z \\
&= \left\| V_2 \right\|_{L^2}^2 \cdot \left\| f \right\|_{L^2}^2 \cdot \left\| \psi \right\|_{L^2}^2 = \left\| V_2 \right\|_{L^2}^2 \cdot \left\| \hat{f} \right\|_{L^2}^2 \cdot \left\| \psi \right\|_{L^2}^2.
\end{aligned}
\tag{1.28}
$$

Choosing $f(x) := (-\Delta + E)^{-1}(x)$, i.e., $\hat{f}(p) = \frac{1}{p^2 + E}$, we observe that

$$\|\hat{f}\|_{L^2}^2 = \int \frac{d^d p}{(p^2 + E)^2} \leq 2 \int \frac{d^d p}{(|p| + E)^4} \leq \frac{C_d}{E^{4-d}}, \tag{1.29}$$

for some constant $C_d < \infty$ depending on d. Thus

$$\lim_{E \to \infty} \|V_2(-\Delta + E)^{-1}\|_{op} \leq \|V_2\|_{L^2} \cdot \lim_{E \to \infty} \|\hat{f}\|_{L^2}^2 \to 0, \tag{1.30}$$

and clearly $\|V_\infty(-\Delta + E)^{-1}\|_{op} \leq \|V_\infty\|_{L^\infty} E^{-1} \to 0$, as well.

- For example, for $d = 3$, $R > 0$, and $V(x) := -|x|^{-1}$, we write

$$V(x) = \frac{-1}{|x|} = \underbrace{\frac{-\mathbf{1}[|x| \leq R]}{|x|}}_{\in L^2(\mathbb{R}^d)} + \underbrace{\frac{-\mathbf{1}[|x| < R]}{|x|}}_{\in L^\infty(\mathbb{R}^d)}. \tag{1.31}$$

Hence $V(x) := -|x|^{-1}$ is an infinitesimal perturbation of $-\Delta$, and $(-\Delta - |x|^{-1}, \mathcal{D}_{-\Delta})$ is self-adjoint and semibounded. In other words, the Kato-Rellich theorem implies the existence of the hydrogen atom and its dynamics.

- The example above shows that relative boundedness works well to prove self-adjointness of *one-body* operators, like the Hamiltonian of the hydrogen atom or, in general, a particle in an external potential. It can also be easily extended to pair potentials, as the following consideration shows. Let $W \in L^2 + L^\infty(\mathbb{R}^3)$ and $\mathcal{H} := L^2(\mathbb{R}^3)$ and define a pair potential operator W_{xy} on $\mathcal{H} \otimes \mathcal{H} \ni \psi$ by

$$[W_{xy}\psi](x, y) := W(x - y)\,\psi(x, y). \tag{1.32}$$

Writing $\psi_y(x) := \psi(x, y)$, we have that for all $a > 0$ there exists $b_a < \infty$ such that

$$\|W_{xy}\psi)\|_{\mathcal{H} \otimes \mathcal{H}}^2 = \int \left(\int |W(x-y)|^2\,|\psi_y(x)|^2\,dx \right) dy$$

$$= \int \|W(\cdot - y)\,\psi_y\|_{\mathcal{H}}^2\,dy \tag{1.33}$$

$$\leq \int \left\|(a(-\Delta) + b_a \cdot \mathbf{1})\,\psi_y\right\|_{\mathcal{H}}^2\,dy$$

$$= a^2 \|(-\Delta \otimes \mathbf{1})\psi\|_{\mathcal{H} \otimes \mathcal{H}}^2 + b_a^2 \|\psi\|_{\mathcal{H} \otimes \mathcal{H}}^2,$$

and similarly

$$\|W_{xy}\psi)\|_{\mathcal{H} \otimes \mathcal{H}}^2 \leq a^2 \|(\mathbf{1} \otimes (-\Delta))\psi\|_{\mathcal{H} \otimes \mathcal{H}}^2 + b_a^2 \|\psi\|_{\mathcal{H} \otimes \mathcal{H}}^2. \tag{1.34}$$

Now we are in position to show self-adjointness and semiboundedness of the Hamiltonian for an atom or a molecule with $N \geq 2$ electrons and $K \geq 1$ fixed nuclei of nuclear charges $\underline{Z} := (Z_1, \ldots, Z_K) \in (\mathbb{R}^+)^K$ located at positions $\underline{R} := (R_1, \ldots, R_K) \in \mathbb{R}^{3K}$. The total nuclear charge is denoted $Z_{tot} := \sum_{k=1}^{K} Z_k$. The Hilbert space of N electrons is

$$\mathcal{H}_A^{(N)} = \mathcal{A}[\mathcal{H}^{(N)}], \tag{1.35}$$

where $\mathcal{H}^{(N)} = \bigotimes_{n=1}^{N} L^2(\mathbb{R}^3 \times \mathbb{Z}_2)$ and $\mathcal{A} = \mathcal{A}^2 = \mathcal{A}^*$ is the projection onto the totally antisymmetric wave functions in accordance with the Pauli principle. The Hamiltonian of the atom $(K = 1)$ or the molecule $(K \geq 2)$ is

$$H_N(\underline{R}, \underline{Z}) = \sum_{n=1}^{N} \left(-\Delta_{x_n} - \sum_{k=1}^{K} \frac{Z_k}{|x_n - R_k|} \right) + \sum_{1 \leq n < m \leq N} \frac{1}{|x_n - x_m|}. \tag{1.36}$$

We may rewrite $H_N \equiv H_N(\underline{R}, \underline{Z})$ as

$$H_N = \mathbb{H}_0 + \mathbb{V}(\underline{x}), \tag{1.37}$$

where $\underline{x} := (x_1, \ldots, x_N) \in \mathbb{R}^{3N}$ and

$$\mathbb{H}_0 := \sum_{n=1}^{N} -\Delta_n, \tag{1.38}$$

$$\mathbb{V}(\underline{x}) := \sum_{n=1}^{N} \sum_{k=1}^{K} \frac{-Z_k}{|x_n - R_k|} + \sum_{1 \leq n < m \leq N} \frac{1}{|x_n - x_m|}. \tag{1.39}$$

Theorem 1.7. *For all $N \geq 2$, $K \geq 1$, $\underline{Z} := (Z_1, \ldots, Z_K) \in (\mathbb{R}^+)^K$, and $\underline{R} := (R_1, \ldots, R_K) \in \mathbb{R}^{3K}$, the Hamiltonian $(H_N, \mathcal{D}_{\mathbb{H}_0} \cap \mathcal{H}^{(N)}) \in \mathcal{L}(\mathcal{H}^{(N)})$ is self-adjoint and semibounded.*

Proof. Since $|x|^{-1}$ is an infinitesimal perturbation of $-\Delta$, we have

$$\||x - R|^{-1}\varphi\| \leq a \cdot \|-\Delta\varphi\| + b \cdot \|\varphi\|, \tag{1.40}$$

for all $\varphi \in \mathcal{D}_{-\Delta}$ and $R \in \mathbb{R}^3$, and thus

$$\left\| \sum_{n=1}^{N} \sum_{k=1}^{K} \frac{Z_k}{|x_n - R_k|} \psi \right\| \leq \sum_{n=1}^{N} \sum_{k=1}^{K} Z_k \left\| \frac{1}{|x_n - R_k|} \psi \right\| \tag{1.41}$$

$$\leq \sum_{n=1}^{N} \sum_{k=1}^{K} Z_k \left(a \|-\Delta_n \psi\| + b \|\psi\| \right)$$

$$= Z_{tot} \left(\sum_{n=1}^{N} a \|-\Delta_n \psi\| + N b \|\psi\| \right).$$

Analogously,

$$\left\| \sum_{n<m} \frac{1}{|x_n - x_m|} \psi \right\| \leq \frac{1}{2} \sum_{n \neq m} \left\| |x_n - x_m|^{-1} \psi \right\| \tag{1.42}$$

$$\leq \frac{1}{2} \sum_{n \neq m} \left\{ \frac{a}{2} \left(\| - \Delta_m \psi \| + \| - \Delta_n \psi \| \right) + b \| \psi \| \right\}$$

$$= \frac{N-1}{2} \left(\sum_{n=1}^{N} a \| - \Delta_n \psi \| \right) + \frac{N(N-1)}{2} b \| \psi \|.$$

Note that, for $m \neq n$,

$$\langle -\Delta_m \psi | -\Delta_n \psi \rangle = \sum_{\mu,\nu=1}^{d} \langle -\partial_{m,\mu}^2 \psi | -\partial_{n,\nu}^2 \psi \rangle$$

$$= \sum_{\mu,\nu=1}^{d} \| \partial_{m,\mu} \partial_{n,\nu} \psi \|^2 \geq 0. \tag{1.43}$$

Hence

$$\sum_{n=1}^{N} \| -\Delta_n \psi \| = \sum_{n=1}^{N} \langle -\Delta_n \psi | -\Delta_n \psi \rangle^{1/2}$$

$$\leq \sum_{m,n=1}^{N} \langle -\Delta_m \psi | -\Delta_n \psi \rangle^{1/2}$$

$$\leq N \left(\sum_{m,n=1}^{N} \langle -\Delta_m \psi | -\Delta_n \psi \rangle \right)^{1/2}$$

$$= N \left\| \sum_{n=1}^{N} -\Delta_n \psi \right\|, \tag{1.44}$$

and we finally obtain

$$\| \mathbb{V}(\underline{x}) \psi \| \leq N \left(Z_{tot} + \frac{N-1}{2} \right) \left(a \| \mathbb{H}_0 \psi \| + b \| \psi \| \right). \tag{1.45}$$

Since $a > 0$ can be chosen arbitrarily small, this implies that $\mathbb{V}(\underline{x})$ is an infinitesimal perturbation of $\mathbb{H}_0$, and the claim follows from Theorem 1.5.

$\square$

Corollary 1.8. *For all $N \geq 2$, $K \geq 1$, $\underline{Z} := (Z_1, \ldots, Z_K) \in (\mathbb{R}^+)^K$, and $\underline{R} := (R_1, \ldots, R_K) \in \mathbb{R}^{3K}$, the Hamiltonian H_N is bounded below by*

$$H_N \geq -8 N^2 \left(Z_{tot} + \frac{N-1}{2} \right)^2. \tag{1.46}$$

Proof. Eq. (1.46) results from an application of (1.25) with $M := 0$, $a := (2L)^{-1}$ which allows us to choose $b_a := 4L$, where $L := N[Z_{tot} + (N-1)/2]$. $\square$

The material of this chapter is covered in various textbooks, for instance, in [13–15] by Reed and Simon.

2. Stability of Matter — Stability of the Second Kind

In this chapter we refine the estimates from Chapter 1 to obtain the lower bound

$$H_N(\underline{Z}, \underline{R}) + U(\underline{Z}, \underline{R}) \geq -C_{SM}(N + K), \tag{2.1}$$

for all N, $\underline{Z}$, and $\underline{R}$, where $C_{SM} \equiv C_{SM}(\zeta)$ only depends on $\zeta := \max\{Z_1, \ldots, Z_K\}$. If such a constant $C_{SM} < \infty$ exists, we call the system **stable of the second kind** or say that **stability of matter** holds true. Stability of matter was proved first by Dyson and Lenard [4, 5]. Later Lieb and Thirring [12] improved this (see [9, 17]) in various aspects.

Definition 2.1. Let $\Psi \in \bigotimes^N L^2(\mathbb{R}^3)$ be normalized. The bounded operator $\gamma_\Psi^{(1)} \in \mathcal{B}[L^2(\mathbb{R}^3)]$ defined by the Schwartz kernel

$$\gamma_\Psi^{(1)}(x, y) := \sum_{n=1}^N \int \prod_{m(\neq n)} d^3 x_m \tag{2.2}$$

$$\left\{ \Psi(x_1, \ldots, x_{n-1}, x, x_{n+1}, \ldots, x_N) \overline{\Psi(x_1, \ldots, x_{n-1}, y, x_{n+1}, \ldots, x_N)} \right\}$$

is called **one-particle density matrix (1-pdm)** of Ψ, and $\rho_\Psi \in L^1(\mathbb{R}^3, \mathbb{R}_0^+)$ given by

$$\rho_\Psi(x) := \sum_{n=1}^N \int \left| \Psi(x_1, \ldots, x_{n-1}, x, x_{n+1}, \ldots, x_N) \right|^2 \prod_{m(\neq n)} d^3 x_m$$

$$= \gamma_\Psi^{(1)}(x, x) \tag{2.3}$$

is called **one-particle density** of Ψ.

Examples and Remarks

- For all normalized $\Psi \in \bigotimes^N L^2(\mathbb{R}^3)$, its 1-pdm $\gamma_\Psi^{(1)}$ is positive and of trace class, $0 \leq \gamma_\Psi^{(1)} \leq \mathrm{Tr}\{\gamma_\Psi^{(1)}\} = N$.
- If $\Psi \in \bigwedge^N L^2(\mathbb{R}^3)$ is a fermion wave function, then the eigenvalues of $\gamma_\Psi^{(1)}$ are even bounded by one, $0 \leq \gamma_\Psi^{(1)} \leq 1$. This easily follows from using the fact that $\langle f | \gamma_\Psi^{(1)} g \rangle = \langle \Psi | c^*(g) c(f) \Psi \rangle$ in this case, where $c^*(g)$ and $c(f)$ are the usual fermion creation and annihilation operators and $\bigwedge^N L^2(\mathbb{R}^3)$ is viewed as the N-particle sector of the fermion Fock space $\mathfrak{F}_f[L^2(\mathbb{R}^3)]$.

Theorem 2.2 ([7, 11]). *There exists a constant $C_{LO} < \infty$ such that for all normalized $\Psi \in \bigotimes^N L^2(\mathbb{R}^3)$,*

$$\left\langle \Psi \middle| \sum_{1 \leq n < m \leq N} \frac{1}{|x_n - x_m|} \Psi \right\rangle \tag{2.4}$$

$$\geq \frac{1}{2} \int \frac{\rho_\Psi(x)\, \rho_\Psi(y)\, d^3x\, d^3y}{|x - y|} - C_{LO} \int \rho_\Psi^{4/3}(x)\, d^3x.$$

Proof. For the proof it is convenient to abbreviate $\rho := \rho_\Psi$. Furthermore, for all $x \neq y$, we use the Fefferman-de la Llave identity

$$\frac{1}{|x - y|} = \frac{1}{\pi} \int d^3z \int_0^\infty \frac{dr}{r^5} \mathbf{1}_{B(z,r)}(x)\, \mathbf{1}_{B(z,r)}(y), \tag{2.5}$$

which yields

$$\sum_{1 \leq n < m \leq N} \frac{1}{|x_n - x_m|} = \frac{1}{2\pi} \int d^3z \int_0^\infty \frac{dr}{r^5} \sum_{m \neq n} \mathbf{1}_{B(z,r)}(x_m)\, \mathbf{1}_{B(z,r)}(x_n)$$

$$= \frac{1}{2\pi} \int d^3z \int_0^\infty \frac{dr}{r^5}\, N_{z,r}(\underline{x})\big[N_{z,r}(\underline{x}) - 1\big], \tag{2.6}$$

with

$$N_{z,r}(\underline{x}) := \sum_{n=1}^N \mathbf{1}_{B(z,r)}(x_n). \tag{2.7}$$

Note that, since $N_{z,r}$ is integer-valued, we have that $N_{z,r}(\underline{x})[N_{z,r}(\underline{x}) - 1] \geq 0$ is nonnegative. Moreover,

$$\langle \Psi | N_{z,r} \Psi \rangle = \sum_{n=1}^N \int \mathbf{1}_{B(z,r)}(x_n)\, |\Psi(x_1, \ldots, x_n, \ldots, x_N)|^2 \prod_{m=1}^N d^3x_m$$

$$= \int_{B(z,r)} \rho(x)\, d^3x, \tag{2.8}$$

and

$$\langle \Psi | \, N_{z,r} \, [N_{z,r}(\underline{x}) - 1] \, \Psi \rangle$$
$$= \langle \Psi | \, N_{z,r}^2 \, \Psi \rangle - \langle \Psi | \, N_{z,r} \, \Psi \rangle$$
$$\geq \langle \Psi | \, N_{z,r} \, \Psi \rangle^2 - \langle \Psi | \, N_{z,r} \, \Psi \rangle$$
$$= \left(\int_{B(z,r)} \rho(x) \, d^3 x \right)^2 - \left(\int_{B(z,r)} \rho(x) \, d^3 x \right), \qquad (2.9)$$

by Jensen's inequality. So, for any measurable choice of $R : \mathbb{R}^3 \to \mathbb{R}_0^+$, we have

$$\left\langle \Psi \left| \sum_{1 \leq n < m \leq N} \frac{1}{|x_n - x_m|} \right| \Psi \right\rangle$$

$$\geq \frac{1}{2\pi} \int d^3 z \int_{R(z)}^{\infty} \frac{dr}{r^5} \, \langle \Psi | \, N_{z,r}(\underline{x}) [N_{z,r}(\underline{x}) - 1] \, \Psi \rangle \qquad (2.10)$$

$$\geq \frac{1}{2\pi} \int d^3 z \int_{R(z)}^{\infty} \frac{dr}{r^5} \left\{ \left(\int_{B(z,r)} \rho(x) \, d^3 x \right)^2 - \left(\int_{B(z,r)} \rho(x) \, d^3 x \right) \right\}.$$

Since by (2.5)

$$\frac{1}{2} \int \frac{\rho(x) \, \rho(y) \, d^3 x \, d^3 y}{|x - y|} = \frac{1}{2\pi} \int d^3 z \int_0^{\infty} \frac{dr}{r^5} \left(\int_{B(z,r)} \rho(x) \, d^3 x \right)^2, \quad (2.11)$$

Eq. (2.10) implies that

$$\left\langle \Psi \left| \sum_{1 \leq n < m \leq N} \frac{1}{|x_n - x_m|} \right| \Psi \right\rangle - \frac{1}{2} \int \frac{\rho(x) \, \rho(y) \, d^3 x \, d^3 y}{|x - y|}$$

$$\geq -\frac{1}{2\pi} \int d^3 z \left\{ \int_0^{R(z)} \left(\int_{B(z,r)} \rho(x) \, d^3 x \right)^2 \frac{dr}{r^5} \right.$$

$$\left. + \int_{R(z)}^{\infty} \left(\int_{B(z,r)} \rho(x) \, d^3 x \right) \frac{dr}{r^5} \right\}. \qquad (2.12)$$

We introduce the *Hardy-Littlewood Maximal Function* $M_\rho : \mathbb{R}^3 \to \mathbb{R}_0^+$ of $\rho \in L^1(\mathbb{R}^3)$ by

$$M_\rho(z) := \sup_{r > 0} \left\{ \frac{1}{|B(z,r)|} \int_{B(z,r)} \rho(x) \, d^3 x \right\}, \qquad (2.13)$$

where $|B(z,r)| := 4\pi r^3/3$ denotes the volume of the ball of radius $r > 0$ about z in $\mathbb{R}^3$. For almost all $z \in \mathbb{R}^3$, we have $\rho(z) \leq M_\rho(z) < \infty$, so $\|\rho\|_{L^p(\mathbb{R}^3)} \leq \|M_\rho\|_{L^p(\mathbb{R}^3)}$, for any $p > 1$. The important nontrivial fact about the Maximal function used here (see [16]) is the *Hardy-Littlewood maximal inequality* which asserts that, for any $p > 1$, the converse inequality $\|M_\rho\|_{L^p(\mathbb{R}^3)} \leq \widehat{C}_p \cdot \|\rho\|_{L^p(\mathbb{R}^3)}$ also holds true, up to multiplication by a constant $\widehat{C}_p < \infty$ which only depends on p (and the spatial dimension which is 3 here). In particular, there exists a universal constant $C' := \widehat{C}_{4/3}^{4/3}$ such that

$$\int_{B(z,r)} M_\rho^{4/3}(x)\, d^3x \leq C' \int_{B(z,r)} \rho^{4/3}(x)\, d^3x. \tag{2.14}$$

Inserting M_ρ into (2.12) and choosing $R(z) := M_\rho^{-1/3}(z)$, we obtain

$$\left\langle \Psi \left| \sum_{1 \leq n < m \leq N} \frac{1}{|x_n - x_m|} \right| \Psi \right\rangle - \frac{1}{2} \int \frac{\rho(x)\,\rho(y)\, d^3x\, d^3y}{|x - y|}$$

$$\geq -\frac{1}{2\pi} \int d^3z \left\{ \frac{16\pi^2}{9} M_\rho^2(z) \left(\int_0^{R(z)} r\, dr \right) + \frac{4\pi}{3} M_\rho(z) \left(\int_{R(z)}^\infty \frac{dr}{r^2} \right) \right\}$$

$$= -\frac{1}{2\pi} \int d^3z \left\{ \frac{8\pi^2}{9} M_\rho^2(z)\, R(z)^2 + \frac{4\pi}{3} M_\rho(z)\, R(z)^{-1} \right\}$$

$$= -C'' \int M_\rho^{4/3}(z)\, d^3z \geq -C'C'' \int \rho^{4/3}(z)\, d^3z, \tag{2.15}$$

where $C'' := (4\pi + 6)/9$. $\qquad\qquad\qquad\qquad\qquad\qquad\qquad\qquad\qquad$ □

Examples and Remarks

- Note that Theorem 2.2 does not assume any antisymmetry of the N-particle wave function ψ, which illustrates the robustness of this estimate.
- On the other hand, the lack of an assumption on the fermionic character of the wave function Ψ also indicates that Theorem 2.2 cannot estimate exchange correlations in the state Ψ accurately.

The second ingredient in our (that is, Lieb's) proof of stability of matter is the Lieb-Thirring inequality.

Theorem 2.3 ([8, 12]). *There exists a constant $C_{LT} < \infty$ such that for all normalized, totally antisymmetric $\Psi \in \bigwedge^N L^2(\mathbb{R}^3)$,*

$$\left\langle \Psi \left| \sum_{n=1}^N -\Delta_n\, \Psi \right\rangle \geq C_{LT} \int \rho_\Psi^{5/3}(x)\, d^3x. \tag{2.16}\right.$$

The Lieb-Oxford inequality (Theorem 2.2) and the Lieb-Thirring inequality
(Theorem 2.3) yield a lower bound on the ground state energy in terms of
the Thomas-Fermi (TF) theory. To see this, we observe that the Cauchy-
Schwarz inequality implies

$$C_{LO} \int \rho_\Psi^{4/3}(x)\, d^3x \le C_{LO} \left(\int \rho_\Psi^{5/3}(x)\, d^3x \right)^{1/2} \left(\int \rho_\Psi(x)\, d^3x \right)^{1/2}$$

$$= C_{LO}\, N^{1/2} \left(\int \rho_\Psi^{5/3}(x)\, d^3x \right)^{1/2}$$

$$\le \frac{C_{LO}^2}{2C_{LT}} N + \frac{C_{LT}}{2} \int \rho_\Psi^{5/3}(x)\, d^3x. \qquad (2.17)$$

Hence we obtain for all normalized, totally antisymmetric $\Psi \in \bigwedge^N L^2(\mathbb{R}^3)$
the lower bound

$$\langle \Psi|\, H_N(\underline{Z}, \underline{R})\, \Psi \rangle \ge E_{C_{LT}/2}^{TF}(N, \underline{Z}, \underline{R}) - \frac{C_{LO}^2}{2C_{LT}} N, \qquad (2.18)$$

where $E_\gamma^{TF}(N, \underline{Z}, \underline{R})$ is the TF energy defined by

$$E_\gamma^{TF}(N, \underline{Z}, \underline{R}) \qquad\qquad\qquad\qquad\qquad\qquad (2.19)$$

$$:= \inf \left\{ \mathcal{E}_{\gamma, \underline{Z}, \underline{R}}^{TF}(\rho) \,\Big|\, \rho \in L^{5/3} \cap L^1,\ \rho \ge 0,\ \int \rho = N \right\}$$

with the TF functional $\mathcal{E}_{\gamma, \underline{Z}, \underline{R}}^{TF} : (L^{5/3} \cap L^1)(\mathbb{R}^3; \mathbb{R}_0^+) \to \mathbb{R}$ is given by

$$\mathcal{E}_{\gamma, \underline{Z}, \underline{R}}^{TF}(\rho) \qquad\qquad\qquad\qquad\qquad\qquad (2.20)$$

$$:= \gamma \int \rho^{5/3}(x)\, d^3x - \sum_{k=1}^{K} \int \frac{Z_k\, \rho(x)\, d^3x}{|x - R_k|} + \frac{1}{2} \int \frac{\rho(x)\, \rho(y)\, d^3x\, d^3y}{|x - y|}.$$

To conclude the proof of stability of matter, we now use three facts from
TF theory.

(i) The TF energy $\mathbb{R}^+ \ni N \mapsto E_\gamma^{TF}(N, \underline{Z}, \underline{R})$ is negative, nonincreasing,
convex function such that

$$\forall N \ge Z_{tot}: \quad E_\gamma^{TF}(N, \underline{Z}, \underline{R}) = E_\gamma^{TF}(Z_{tot}, \underline{Z}, \underline{R}), \qquad (2.21)$$

where $Z_{tot} := \sum_{k=1}^{N} Z_k$ is the total nuclear charge of the system.

(ii) According to Teller's lemma, atoms do not bind to molecules in TF theory. More precisely,

$$\inf \left\{ E_\gamma^{TF}(Z_{tot}, \underline{Z}, \underline{R}) + U(\underline{Z}, \underline{R}) \,\Big|\, \underline{R} \in (\mathbb{R}^3)^K \right\} \tag{2.22}$$

$$= \lim_{r \to \infty} E_\gamma^{TF}(Z_{tot}, \underline{Z}, r\underline{R}) = \sum_{k=1}^{K} E_\gamma^{TF}(Z_k, Z_k, 0),$$

where the right side is the sum of the TF energies of K single neutral atoms of nuclear charges $Z_1, \ldots, Z_K$.

(iii) By scaling, one sees that the TF energy of a neutral atom of nuclear charge Z is given by

$$E_\gamma^{TF}(Z, Z, 0) = -C_{TF}(\gamma) \, Z^{7/3}, \tag{2.23}$$

where $C_{TF}(\gamma) := -E_\gamma^{TF}(1, 1, 0)$ is the TF energy of the hydrogen atom.

Combining (i)–(iii) with (2.18), we finally arrive at

Theorem 2.4 ([12]). *For all $N \in \mathbb{N}$, all normalized, totally antisymmetric $\Psi \in \bigwedge^N L^2(\mathbb{R}^3)$ and all $\underline{R} \in (\mathbb{R}^3)^K$, the ground state energy obeys the following lower bound*

$$\langle \Psi | \, H_N(\underline{Z}, \underline{R}) \, \Psi \rangle + U(\underline{Z}, \underline{R}) \geq - C_{SM}(N + K), \tag{2.24}$$

where $C_{SM} := C_{TF}(C_{LT}/2) \max_k Z_k^{7/3} + C_{LO}^2 (2C_{LT})^{-1} < \infty$ only depends on the maximal nuclear charge $\max_k Z_k$.

Examples and Remarks

• In the limit of large Z, the ground state energy

$$E_{\text{gs}}(N, \underline{Z}, \underline{R}) := \inf \left\{ \langle \Psi | \, H_N(\underline{Z}, \underline{R}) \, \Psi \rangle \,\Big|\, \Psi \in \bigwedge^N \mathfrak{h} \cap \otimes^N \mathfrak{d}, \ \|\Psi\| = 1 \right\}, \tag{2.25}$$

of an atom or molecule scales like $Z^{7/3}$. The large-Z limit is defined by $Z_{tot} \to \infty$ under the assumption of a fixed number of nuclei and mutual ratios of the nuclear charges and the number of electrons comparable to the total nuclear charge. More precisley, let $\underline{z} = (z_1, \ldots, z_K) \in (\mathbb{R}^+)^K$ be fixed such that $z_1 + \ldots + z_K = 1$ and set $\underline{Z} := Z \cdot \underline{z}$, where $Z > 0$. Furthermore, we assume that $\frac{1}{2} Z \leq N \leq 2Z$. Then there are constants $0 < C_1 < C_2 < \infty$ such that

$$-C_2 \, K \, Z^{7/3} \leq E_{\text{gs}}(N, \underline{Z}, \underline{R}) \leq -C_1 \, Z^{7/3}, \tag{2.26}$$

- Moreover, there exist constants $0 < C_1' < C_2' < \infty$ such that, if $\Psi \in \bigwedge^N L^2(\mathbb{R}^3)$, with $\|\Psi\| = 1$, is a very rough approximation to a ground state, i.e.,

$$E_{\mathrm{gs}}(N, \underline{Z}, \underline{R}) \leq \langle \Psi | \, H_N(\underline{Z}, \underline{R}) \, \Psi \rangle \leq \frac{1}{2} E_{\mathrm{gs}}(N, \underline{Z}, \underline{R}), \qquad (2.27)$$

then also each of its contributions to its energy expectation value is of order $Z^{7/3}$,

$$C_1' \, Z^{7/3} \leq \int \rho_\Psi^{5/3}(x) \, d^3x \leq C_2' \, Z^{7/3}, \qquad (2.28)$$

$$C_1' \, Z^{7/3} \leq \sum_{k=1}^{K} \int \frac{Z \, z_k \, \rho_\Psi(x) \, d^3x}{|x - R_k|} \leq C_2' \, Z^{7/3}, \qquad (2.29)$$

$$C_1' \, Z^{7/3} \leq \frac{1}{2} \int \frac{\rho(x) \, \rho(y) \, d^3x \, d^3y}{|x - y|} \leq C_2' \, Z^{7/3}. \qquad (2.30)$$

- The exchange integral, however, is much smaller, namely

$$\int \rho_\Psi^{4/3}(x) \, d^3x \leq \left(\int \rho_\Psi^{5/3}(x) \, d^3x \right)^{1/2} \left(\int \rho_\Psi(x) \, d^3x \right)^{1/2}$$

$$\leq C_2'' \, Z^{5/3}. \qquad (2.31)$$

3. Hartree-Fock (HF) Theory for Atoms and Molecules

It is convenient to follow [3] and introduce the fermion Fock space $\mathfrak{F}[\mathfrak{h}]$ over a one-particle Hilbert space $\mathfrak{h}$ by $\mathfrak{F}^{(0)}[\mathfrak{h}] := \mathbb{C} \cdot \Omega$, where Ω is normalized and called vaccum vector, and

$$\mathfrak{F}^{(N)}[\mathfrak{h}] := \bigwedge^N \mathfrak{h}, \qquad \mathfrak{F}[\mathfrak{h}] := \bigoplus_{N=0}^{\infty} \mathfrak{F}^{(N)}[\mathfrak{h}], \qquad (3.1)$$

so, for $\mathfrak{h} = L^2(\mathbb{R}^3)$, we have $\mathfrak{F}^{(N)}[\mathfrak{h}] = \mathcal{H}^{(N)}$, as before. We use creation and annilhilation operators, $c^*(f)$ and $c(g)$, defined by $c(g) := \left(c^*(g) \right)^*$ and

$$c^*(\varphi_1) \big[c^*(\varphi_2) \cdots c^*(\varphi_N) \, \Omega \big]$$
$$:= c^*(\varphi_1) \, c^*(\varphi_2) \cdots c^*(\varphi_N) \, \Omega$$
$$:= \varphi_1 \wedge \varphi_2 \wedge \cdots \wedge \varphi_N$$
$$:= (N!)^{-1/2} \sum_{\pi \in \mathcal{S}_N} (-1)^\pi \varphi_{\pi(1)} \otimes \varphi_{\pi(2)} \otimes \cdots \otimes \varphi_{\pi(N)}. \qquad (3.2)$$

The operator family $\{c^*(f), c(f)\}_{f \in \mathfrak{h}} \subseteq \mathcal{B}(\mathfrak{F}[\mathfrak{h}])$ is a *Fock representation of the canonical anticommutation relation (CAR)*, i.e., for all $f, g \in \mathfrak{h}$, we have

$$\{c(f), c^*(g)\} = \langle f | g \rangle \, \mathbf{1}, \tag{3.3}$$

$$\{c(f), c(g)\} = \{c^*(f), c^*(g)\} = 0, \tag{3.4}$$

$$c(g) \, \Omega = 0. \tag{3.5}$$

Note that

$$\left\| c^*(f) \right\|_{\mathrm{op}} = \left\| c(f) \right\|_{\mathrm{op}} = \|f\|_{\mathfrak{h}}, \tag{3.6}$$

because $c^*(f)\Omega = f$ and

$$\left\| c^*(f) \, \Psi \right\|_{\mathrm{op}}^2 + \left\| c(f) \, \Psi \right\|_{\mathrm{op}}^2 = \langle \Psi \, | \, \{c(f), c^*(f)\} \, \Psi \rangle = \|f\|_{\mathfrak{h}}^2 \cdot \|\Psi\|^2, \tag{3.7}$$

for any $\Psi \in \mathfrak{F}[\mathfrak{h}]$, thanks to (3.3).

Now let $\{\varphi_\alpha\}_{\alpha=1}^\infty \subseteq \mathfrak{h}$ be an ONB of sufficiently regular vectors (e.g., eigenfunctions of the harmonic oscillator, in case that $\mathfrak{h} = L^2(\mathbb{R}^3)$). An N-particle Hamiltonian of the form

$$H_N = \sum_{n=1}^N h_n + \sum_{1 \leq m < n \leq N} V_{m,n}, \tag{3.8}$$

with $h_n = \mathbf{1}^{\otimes(n-1)} \otimes h \otimes \mathbf{1}^{\otimes(N-n)}$ and, e.g., $h = -\Delta_x - \sum_{k=1}^K Z_k |x - R_k|^{-1}$, as well as $V_{m,n} = V(x_m - x_n)$, can be viewed as the restriction of

$$\mathbb{H} = \sum_{\alpha,\beta=1}^\infty h_{\alpha,\beta} \, c_\alpha^* \, c_\beta + \sum_{\alpha,\beta,\tilde{\alpha},\tilde{\beta}=1}^\infty V_{\alpha,\tilde{\alpha};\beta,\tilde{\beta}} \, c_{\tilde{\alpha}}^* \, c_\alpha^* \, c_\beta \, c_{\tilde{\beta}}, \tag{3.9}$$

to $\mathfrak{F}^{(N)}[\mathfrak{h}]$, where

$$h_{\alpha,\beta} := \langle \varphi_\alpha | \, h \, \varphi_\beta \rangle, \tag{3.10}$$

$$V_{\alpha,\tilde{\alpha};\beta,\tilde{\beta}} := \langle \varphi_\alpha \otimes \varphi_{\tilde{\alpha}} | \, V \, (\varphi_\beta \otimes \varphi_{\tilde{\beta}}), \tag{3.11}$$

$$c_\alpha^* = c^*(\varphi_\alpha), \quad c_\alpha = c(\varphi_\alpha). \tag{3.12}$$

Definition 3.1. Let $\mathfrak{h}$ be a Hilbert space.

(i) A **density matrix** is a positive operator $\rho \in \mathcal{B}(\mathfrak{F}[\mathfrak{h}])$ of unit trace, $0 \leq \rho \leq \mathrm{Tr}_{\mathfrak{F}}\{\rho\} = 1$.

(ii) If $\rho \in \mathcal{B}(\mathfrak{F}[\mathfrak{h}])$ is a density matrix, its **one-particle density matrix (1-pdm)** $\gamma_\rho^{(1)} \in \mathcal{B}(\mathfrak{h})$ is determined by

$$\langle f \, | \, \gamma_\rho^{(1)} \, g \rangle_{\mathfrak{h}} := \mathrm{Tr}_{\mathfrak{F}}\{\rho \, c^*(g) \, c(f)\}. \tag{3.13}$$

(iii) If $\rho \in \mathcal{B}(\mathfrak{F}[\mathfrak{h}])$ is a density matrix, its **two-particle density matrix (2-pdm)** $\gamma_\rho^{(2)} \in \mathcal{B}(\mathfrak{h} \otimes \mathfrak{h})$ is determined by

$$\langle f \otimes \tilde{f} \mid \gamma_\rho^{(2)} (g \otimes \tilde{g}) \rangle_{\mathfrak{h} \otimes \mathfrak{h}} := \mathrm{Tr}_{\mathfrak{F}} \{ \rho \, c^*(g) \, c^*(\tilde{g}) \, c(\tilde{f}) \, c(f) \}. \quad (3.14)$$

Obviously, we have

$$\langle \Psi \mid \mathbb{H} \Psi \rangle_{\mathfrak{F}} = \mathrm{Tr}_{\mathfrak{h}} \{ h \, \gamma_\rho^{(1)} \} + \frac{1}{2} \mathrm{Tr}_{\mathfrak{h} \otimes \mathfrak{h}} \{ V \, \gamma_\rho^{(2)} \}, \quad (3.15)$$

where $\rho = |\Psi\rangle\langle\Psi|$.

Lemma 3.2. *Let* $\rho \in \mathcal{B}(\mathfrak{F}[\mathfrak{h}])$ *be a density matrix such that* $\mathrm{Tr}\{\rho \mathbb{N}^2\} < \infty$, *where*

$$\mathbb{N} := \sum_{\alpha=1}^{\infty} c_\alpha^* c_\alpha = \bigoplus_{N=0}^{\infty} N \cdot \mathbf{1}_{\mathfrak{F}^{(N)}[\mathfrak{h}]} \quad (3.16)$$

is the number operator on $\mathfrak{F}[\mathfrak{h}]$. *Then* $\gamma_\rho^{(1)} \in \mathcal{L}^1[\mathfrak{h}]$ *and* $\gamma_\rho^{(2)} \in \mathcal{L}^1[\mathfrak{h} \otimes \mathfrak{h}]$ *are trace-class and*

$$0 \leq \gamma_\rho^{(1)} \leq \mathbf{1}, \qquad\qquad \mathrm{Tr}_{\mathfrak{h}} \{ \gamma_\rho^{(1)} \} = \mathrm{Tr}_{\mathfrak{F}} \{ \rho \, \mathbb{N} \}, \quad (3.17)$$

$$0 \leq \gamma_\rho^{(2)} \leq \mathrm{Tr}_{\mathfrak{F}} \{ \rho \, \mathbb{N} \} \cdot \mathbf{1}, \quad \mathrm{Tr}_{\mathfrak{h} \otimes \mathfrak{h}} \{ \gamma_\rho^{(2)} \} = \mathrm{Tr}_{\mathfrak{F}} \{ \rho \, \mathbb{N}(\mathbb{N} - 1) \}. \quad (3.18)$$

Moreover, $\gamma_\rho^{(1)}$ *is a rank-N orthogonal projection onto orthonormal orbitals* $\varphi_1, \ldots, \varphi_N \in \mathfrak{h}$ *if, and only if,* $\rho = |\Phi\rangle\langle\Phi|$ *is a rank-1 projection onto the Slater determinant* $\Phi = \varphi_1 \wedge \ldots \wedge \varphi_N$, *i.e.,*

$$\left\{ \gamma_\rho^{(1)} = \sum_{n=1}^{N} |\varphi_n\rangle\langle\varphi_n| = (\gamma_\rho^{(1)})^2 \right\} \iff \{ \rho = |\Phi\rangle\langle\Phi|, \ \Phi = \varphi_1 \wedge \ldots \wedge \varphi_N \}. \quad (3.19)$$

Examples and Remarks

- In general, $\gamma_\rho^{(1)} - (\gamma_\rho^{(1)})^2 \geq 0$.
- If (3.19) holds true, that is, if $\gamma_\rho^{(1)}$ is a rank-N projection, then also $\frac{1}{2}\gamma_\rho^{(2)}$ is a projection of rank $\frac{1}{2}N(N-1)$, namely,

$$\gamma_\rho^{(2)} = \sum_{m,n=1}^{N} |\varphi_m \wedge \varphi_n\rangle\langle\varphi_m \wedge \varphi_n| = \gamma_\rho^{(1)} \otimes \gamma_\rho^{(1)} - \mathrm{Ex}(\gamma_\rho^{(1)} \otimes \gamma_\rho^{(1)}), \quad (3.20)$$

where $\mathrm{Ex}(\varphi \otimes \psi) := \psi \otimes \varphi$ is the **exchange operator**.

Definition 3.3. Let $(h, \mathfrak{d}) \in \mathcal{L}(\mathfrak{h})$ be self-adjoint and semibounded, $h \geq -C + 1$, and $(V, \mathfrak{d} \otimes \mathfrak{d}) \in \mathcal{L}(\mathfrak{h} \otimes \mathfrak{h})$ symmetric, positive, $[V, \text{Ex}] = 0$, and obeying

$$0 \leq V \leq \frac{1}{4}\left(h \otimes \mathbf{1} + \mathbf{1} \otimes h\right) + C. \tag{3.21}$$

(i) The **ground state energy** $E_{\text{gs}}(N)$ **(for $N \in \mathbb{N}$ particles)** is defined by

$$E_{\text{gs}}(N) := \inf\left\{\langle\Psi|\mathbb{H}\Psi\rangle \;\middle|\; \Psi \in \overset{N}{\bigwedge}\mathfrak{h} \cap \otimes^N\mathfrak{d}, \; \|\Psi\| = 1\right\}. \tag{3.22}$$

(ii) The **Hartree-Fock (HF) energy** $E_{\text{hf}}(N)$ **(for $N \in \mathbb{N}$ particles)** is defined by

$$E_{\text{hf}}(N) := \inf\left\{\langle\Phi|\mathbb{H}\Phi\rangle \;\middle|\; \Phi = c^*(\varphi_1)\cdots c^*(\varphi_N)\Omega, \right. \tag{3.23}$$
$$\left.\varphi_n \in \mathfrak{d}, \; \langle\varphi_m|\varphi_n\rangle = \delta_{m,n}\right\}.$$

(iii) For $\gamma \in \mathcal{L}^1(\mathfrak{h})$, $0 \leq \gamma \leq 1$, $\text{Tr}\{\gamma\} < \infty$, $\text{Tr}\{h\gamma\} < \infty$, the **HF functional** is given by

$$\mathcal{E}_{\text{hf}}(\gamma) := \text{Tr}_{\mathfrak{h}}\{h\,\gamma\} + \frac{1}{2}\text{Tr}_{\mathfrak{h}\otimes\mathfrak{h}}\{V\,(1 - \text{Ex})\,(\gamma \otimes \gamma)\}. \tag{3.24}$$

Examples and Remarks

- Since the variation in (3.23) takes places over a smaller set as compared to (3.22), clearly $E_{\text{hf}}(N) \geq E_{\text{gs}}(N)$ holds true.
- Eq. (3.19) implies that

$$E_{\text{hf}}(N) = \inf\left\{\mathcal{E}_{\text{hf}}(\gamma) \,\middle|\, \gamma = \gamma^* = \gamma^2, \; \text{Tr}(\gamma) = N, \; \text{Tr}\{h\gamma\} < \infty\right\}. \tag{3.25}$$

- For $\mathfrak{h} = L^2(\mathbb{R}^3)$, $h = -\Delta_x - \frac{Z}{|x|}$, $V_{xy} = \frac{1}{|x-y|}$, and $Z/2 \leq N \leq 2Z$, we have

$$\mathcal{E}_{\text{hf}}(\gamma) = \text{Tr}\left\{\left(-\Delta - \frac{z}{|x|}\right)\gamma\right\} + \frac{1}{2}\int \frac{\rho(x)\,\rho(y) - |\gamma(x,y)|^2}{|x - y|}d^3x\,d^3y. \tag{3.26}$$

Lemma 3.4 (Lieb's Variational Principle [10]). *Assume the hypothesis of Definition 3.3, in particular, $V \geq 0$. Then*

$$E_{\text{hf}}(N) = \inf\left\{\mathcal{E}_{\text{hf}}(\gamma) \,\middle|\, 0 \leq \gamma \leq 1, \;\; \text{Tr}(\gamma) = N, \;\; \text{Tr}\{h\gamma\} < \infty\right\}. \tag{3.27}$$

Proof ([1]). Let $\gamma = \sum_{n=1}^{M} \lambda_n |\varphi_n\rangle\langle\varphi_n|$, $\langle\varphi_m|\varphi_n\rangle = \delta_{m,n}$, $0 < \lambda_n \leq 1$, $\sum_{n=1}^{M} \lambda_n = N$. Assume $M > N$ and, for the sake of simplicity, $M < \infty$. Denote

$$h_n := \langle\varphi_n|h\varphi_n\rangle, \quad V_{m,n} := \langle\varphi_n \wedge \varphi_n | V(\varphi_m \wedge \varphi_n)\rangle \geq 0, \qquad (3.28)$$

and

$$p_n := h_n + \sum_{m=1}^{M} \lambda_m V_{m,n} \leq p_{n+1}. \qquad (3.29)$$

Then

$$\mathcal{E}_{\mathrm{hf}}(\gamma) = \sum_{n=1}^{M} \lambda_n h_n + \frac{1}{2} \sum_{m,n=1}^{M} V_{m,n} \lambda_m \lambda_n. \qquad (3.30)$$

Since $M > N$,

$$\exists 1 \leq k < \ell \leq M : \quad 0 < \lambda_k, \lambda_\ell < 1. \qquad (3.31)$$

We fix k, ℓ according to (3.31) and define γ_δ by

$$\gamma_\delta := \gamma + \delta\Big(|\varphi_k\rangle\langle\varphi_k| - |\varphi_\ell\rangle\langle\varphi_\ell|\Big). \qquad (3.32)$$

Then $\mathrm{Tr}\{\gamma_\delta\} = N$ and $0 \leq \gamma_\delta \leq 1$, provided $\delta > 0$ is chosen sufficiently small. Furthermore,

$$\begin{aligned}
\mathcal{E}_{\mathrm{hf}}&(\gamma_\delta) - \mathcal{E}_{\mathrm{hf}}(\gamma) \hspace{4cm} (3.33)\\
&= \delta(h_k - h_\ell)\\
&\quad + \sum_{m,n=1}^{M} V_{m,n}\{(\lambda_m + \delta(\delta_{m,k} - \delta_{m,\ell}))(\lambda_n + \delta(\delta_{n,k} - \delta_{n,\ell})) - \lambda_m \lambda_n\}\\
&= \delta(p_k - p_\ell) - \delta^2 V_{k,\ell} < 0,
\end{aligned}$$

using $\delta > 0$, $p_k - p_\ell \leq 0$, $V_{k,\ell} > 0$, and $V_{k,k} = V_{\ell,\ell} = 0$. Choosing $\delta := \min\{1 - \lambda_k, \lambda_\ell\} > 0$, we obtain

$$\mathrm{rk}(\gamma_\delta) \leq \mathrm{rk}(\gamma) - 1. \qquad (3.34)$$

After at most $M - N$ such steps, we arrive at a rank-N projection. $\qquad\square$

In the following, we again assume that

$$\mathfrak{h} := L^2(\mathbb{R}^3), \quad h = -\Delta_x - \frac{Z}{|x|}, \tag{3.35}$$

$$V(x-y) := \frac{1}{|x-y|} = \frac{1}{\pi} \int d^3z \int_0^\infty \frac{dr}{r^5} \mathbf{1}_{B(z,r)}(x) \mathbf{1}_{B(z,r)}(y). \tag{3.36}$$

To show that $E_{\mathrm{hf}}(N)$ agrees with $E_{\mathrm{gs}}(N)$ to high accuracy, we need to improve the Lieb-Oxford inequality

$$\left\langle \Psi \,\middle|\, \sum_{m<n} |x_m - x_n|^{-1} \Psi \right\rangle$$

$$\geq \frac{1}{2} \int \rho_\Psi(x)\, \rho_\Psi(y)\, \frac{d^3x\, d^3y}{|x-y|} - C_{LO} \int \rho_\Psi^{4/3}(x)\, d^3x. \tag{3.37}$$

We have remarked in (2.26)–(2.31) that, among all contributions to the ground state energy of a large atom, the exchange correction is the smallest in magnitude and can be estimated by $Z^{5/3}$, according to (2.31). So, an agreement of $E_{\mathrm{hf}}(N)$ and $E_{\mathrm{gs}}(N)$ to high accuracy means that, the difference $E_{\mathrm{hf}}(N) - E_{\mathrm{gs}}(N)$ is small compared to $Z^{5/3}$. Such an estimate of the accuray is precisely the contents of the next theorem found in [1].

Theorem 3.5. *Let $\underline{z} = (z_1, \ldots, z_K) \in (\mathbb{R}^+)^K$ be fixed such that $z_1 + \ldots + z_K = 1$ and set $\underline{Z} := Z \cdot \underline{z}$, where $Z > 0$. Furthermore, assume that $\frac{1}{2}Z \leq N \leq 2Z$. Then there exists a constant $C < \infty$ such that*

$$0 \leq E_{\mathrm{hf}}(N) - E_{\mathrm{gs}}(N) \leq C\, Z^{5/3 - 1/7}, \tag{3.38}$$

The heart of the proof of Theorem 3.5 is the following correlation estimate from [1]:

Lemma 3.6 (Fermion Correlation Estimate). *For any orthogonal projection $X = X^2 = X^*$ and any particle number-conserving density matrix $\rho \in \mathcal{B}(\mathfrak{F}[\mathfrak{h}])$, we have that*

$$\mathrm{Tr}\left\{ (X \otimes X)\left[\gamma_\rho^{(2)} - (1 - \mathrm{Ex})(\gamma_\rho^{(1)} \otimes \gamma_\rho^{(1)}) \right] \right\}$$

$$\geq -\mathrm{Tr}\{X\gamma_\rho^{(1)}\} \cdot \sqrt{\min\left(1,\, 8\,\mathrm{Tr}\{X(\gamma_\rho - \gamma_\rho^2)\}\right)}. \tag{3.39}$$

Using (3.39) with $X := \mathbf{1}_{B(z,r)}$, we obtain

$$
\left\langle \Psi \,\middle|\, \sum_{m<n} |x_m - x_n|^{-1} \Psi \right\rangle - \frac{1}{2} \int \left(\rho_\Psi(x)\, \rho_\Psi(y) - |\gamma_\Psi^{(1)}(x,y)|^2 \right) \frac{d^3x\, d^3y}{|x-y|}
$$

$$
\geq -\frac{5}{\pi} \int d^3z \left\{ \int_0^{R(z)} \left(\int_{B(z,r)} \rho_\Psi(x)\, d^3x \right)^2 \frac{dr}{r^5} \right.
$$

$$
\left. + \int_{R(z)}^\infty \left(\mathrm{Tr}\{\mathbf{1}_{B(z,r)}(\gamma_\rho - \gamma_\rho^2)\} \right) \frac{dr}{r^5} \right\}, \tag{3.40}
$$

as in (2.12), and further proceeding as in the proof of the Lieb-Oxford inequality, we indeed obtain

Lemma 3.7. *There is a constant $C < \infty$ such that for all normalized $\Psi \in \bigwedge^N \mathfrak{h} \cap \bigotimes^N \mathfrak{d}$, we have*

$$
\left\langle \Psi \,\middle|\, \sum_{m<n} |x_m - x_n|^{-1} \Psi \right\rangle
$$

$$
\geq \frac{1}{2} \int \left(\rho_\Psi(x)\, \rho_\Psi(y) - |\gamma_\Psi^{(1)}(x,y)|^2 \right) \frac{d^3x\, d^3y}{|x-y|}
$$

$$
- C \left(\int \rho_\Psi^{4/3}(x)\, d^3x \right) \left(\frac{\mathrm{Tr}\{\gamma_\Psi - \gamma_\Psi^2\}}{\mathrm{Tr}\{\gamma_\Psi\}} \right)^{1/3}. \tag{3.41}
$$

The final ingredient of the proof is the following estimate which states that if $\Psi \in \bigwedge^N \mathfrak{h} \cap \bigotimes^N \mathfrak{d}$ is (very close to) a ground state of the system then γ_Ψ cannot deviate much from an orthogonal projection - much like the projection onto the Fermi sea. This follows from semiclassical estimates by Ivrii and Sigal [6].

Lemma 3.8. *Let $\underline{z} = (z_1, \ldots, z_K) \in (\mathbb{R}^+)^K$ be fixed such that $z_1 + \cdots + z_K = 1$ and set $\underline{Z} := Z \cdot \underline{z}$, where $Z > 0$. Furthermore, assume that $\frac{1}{2}Z \leq N \leq 2Z$. Then there exists a constant $C < \infty$ such that, for any normalized ground state $\Psi \in \bigwedge^N \mathfrak{h} \cap \bigotimes^N \mathfrak{d}$, i.e., $\langle \Psi | \mathbb{H}\Psi \rangle = E_{\mathrm{gs}}(N)$, we have that*

$$
\mathrm{Tr}\{\gamma_\Psi - \gamma_\Psi^2\} \leq C\, Z^{-3/7}\, \mathrm{Tr}\{\gamma_\Psi\}. \tag{3.42}
$$

Proof (Sketch of the Proof of Theorem 3.5). Inserting (3.42) into (3.41), we obtain a the following lower bound for the pair interaction energy of a normalized ground state $\Psi \in \bigwedge^N \mathfrak{h} \cap \bigotimes^N \mathfrak{d}$,

$$\left\langle \Psi \,\middle|\, \sum_{m<n} |x_m - x_n|^{-1} \Psi \right\rangle$$

$$\geq \frac{1}{2} \int \left(\rho_\Psi(x)\, \rho_\Psi(y) - \left| \gamma_\Psi^{(1)}(x,y) \right|^2 \right) \frac{d^3x\, d^3y}{|x-y|} - C\, Z^{5/3 - 1/7}. \quad (3.43)$$

Therefore,

$$\langle \Psi \,|\, \mathbb{H}\, \Psi \rangle \geq \mathcal{E}_{\mathrm{hf}}\big(\gamma_\Psi^{(1)}\big) - C\, Z^{5/3-1/7}, \quad (3.44)$$

from which the assertion is immediate. □

4. Bogolubov-Hartree-Fock (BHF) Theory

The goal of this section is to extend HF theory from a variation over Slater determinants to a variation over quasifree density matrices.

Definition 4.1. Let $\mathfrak{h} = L^2(M)$ be a Hilbert space given as a space of square-integrabel functions on a measure space M, and denote $\bar{f}(x) := \overline{f(x)}$, for all $f \in \mathfrak{h}$ and $x \in M$. Let $\rho \in \mathcal{B}(\mathfrak{F}[\mathfrak{h}])$ be a density matrix.

(i) The **generalized one-particle density matrix (g1-pdm)** $\Gamma_\rho \in \mathcal{B}[\mathfrak{h} \oplus \mathfrak{h}]$ corresponding to ρ is defined by

$$\left\langle \begin{pmatrix} f_1 \\ f_2 \end{pmatrix} \,\middle|\, \Gamma_\rho \begin{pmatrix} g_1 \\ g_2 \end{pmatrix} \right\rangle := \mathrm{Tr}\big\{ \rho\, \big([c^*(g_1) + c(\bar{g}_2)]\, [c^*(\bar{f}_1) + c(f_2)] \big) \big\}, \quad (4.1)$$

(ii) The **pairing matrix** $\alpha_\rho \in \mathcal{B}[\mathfrak{h}]$ corresponding to ρ is defined by

$$\langle f | \alpha_\rho\, g \rangle := \mathrm{Tr}\big\{ \rho\, c(\bar{g})\, c(f) \big\}. \quad (4.2)$$

(iii) The density matrix ρ is called **quasifree** iff, given $2n \in 2\mathbb{N}$ orbitals $f_1, \ldots, f_{2n} \in \mathfrak{h}$ and $2n$ creation and annihilation operators $C_1, \ldots, C_{2n} \in \{c^*(f_1), c(f_1), \ldots, c^*(f_{2n}),\, c(f_{2n})\}$, we have $\mathrm{Tr}\{\rho\, C_1 \cdots C_{2n-1}\} = 0$ and

$$\mathrm{Tr}\{\rho\, C_1 \cdots C_{2n}\} \quad (4.3)$$

$$= \sum_{\pi \in \mathcal{P}_{2n}} (-1)^\pi\, \mathrm{Tr}\{\rho\, C_{\pi(1)} C_{\pi(2)}\} \cdots \mathrm{Tr}\{\rho\, C_{\pi(2n-1)} C_{\pi(2n)}\},$$

where $\mathcal{P}_{2n} \subseteq \mathcal{S}_{2n}$ denotes the set of pairings of $2n$ elements, i.e., all permutations $\pi \in \mathcal{S}_{2n}$, for which $\pi(1) < \pi(3) < \cdots < \pi(2n-1)$ and $\pi(1) < \pi(2),\, \pi(3) < \pi(4), \ldots, \pi(2n-1) < \pi(2n)$.

Examples and Remarks

- Due to the CAR, we obviously have

$$0 \leq \Gamma_\rho \leq 1, \tag{4.4}$$

which is equivalent to

$$\Gamma_\rho - \Gamma_\rho^2 \geq 0. \tag{4.5}$$

- The g1-pdm Γ_ρ can be conveniently expressed in terms of a 2×2-matrix $\Gamma_\rho = \Gamma[\gamma_\rho, \alpha_\rho]$ whose entries involve the 1-pdm $\gamma_\rho^{(1)}$ and the pairing matrix α_ρ, where

$$\Gamma[\gamma, \alpha] = \begin{pmatrix} \gamma & \alpha \\ \alpha^* & 1 - \overline{\gamma} \end{pmatrix}, \tag{4.6}$$

- While $\gamma_\rho^{(1)}$ is self-adjoint and positive, α_ρ is antisymmetric,

$$\gamma_\rho^{(1)} = (\gamma_\rho^{(1)})^*, \qquad \alpha_\rho = -\alpha_\rho^T, \tag{4.7}$$

where $\alpha_\rho^T(x, y) := \alpha_\rho(y, x)$, for $x, y \in M$, denotes the transpose of α_ρ.
- Eq. (4.5) implies that

$$\gamma_\rho^{(1)} \geq (\gamma_\rho^{(1)})^2 + \alpha_\rho \alpha_\rho^*. \tag{4.8}$$

- If ρ commutes with the particle number operator, in particular, if $\rho = |\Phi\rangle\langle\Phi|$ is the state corresponding to a Slater determinant Φ then its pairing matrix vanishes. In this case, (4.8) reduces to $0 \leq \gamma_\rho^{(1)} \leq 1$
- Quasifree density matrices and generalized one-particle density matrices are in one-to-one correspondence to another. More specifically:

$$\rho \text{ density matrix}, \ \mathrm{Tr}\{\rho\mathbb{N}\} < \infty \implies \tag{4.9}$$
$$0 \leq \Gamma[\gamma_\rho^{(1)}, \alpha_\rho] \leq 1, \quad \gamma_\rho^{(1)} = (\gamma_\rho^{(1)})^*,$$
$$\alpha_\rho = -\alpha_\rho^T, \quad \mathrm{Tr}\{\gamma_\rho^{(1)}\} = \mathrm{Tr}\{\rho\mathbb{N}\},$$

and conversely,

$$0 \leq \Gamma := \Gamma[\gamma, \alpha] \leq 1, \quad \gamma = \gamma^*, \quad \alpha = -\alpha^T, \quad \mathrm{Tr}\{\gamma\} < \infty, \tag{4.10}$$

$\implies$ There exists a quasifree density matrix ρ such that $\Gamma_\rho = \Gamma$.

- The case $n = 2$ in (4.3) is of special interest. Namely, if ρ is quasifree, then

$$\mathrm{Tr}\{\rho\, C_1 C_2 C_3 C_4\}$$
$$= \mathrm{Tr}\{\rho\, C_1 C_4\}\,\mathrm{Tr}\{\rho\, C_2 C_3\}$$
$$- \mathrm{Tr}\{\rho\, C_1 C_3\}\,\mathrm{Tr}\{\rho\, C_2 C_4\} + \mathrm{Tr}\{\rho\, C_1 C_2\}\,\mathrm{Tr}\{\rho\, C_3 C_4\}, \tag{4.11}$$

which yields

$$\langle f \otimes \tilde{f} \mid \gamma_\rho^{(2)} (g \otimes \tilde{g}) \rangle$$
$$= \langle f \mid \gamma_\rho^{(1)} \tilde{g} \rangle \langle \tilde{f} \mid \gamma_\rho^{(1)} g \rangle$$
$$- \langle f \mid \gamma_\rho^{(1)} g \rangle \langle \tilde{f} \mid \gamma_\rho^{(1)} \tilde{g} \rangle + \langle \tilde{g} \mid \alpha_\rho^* g \rangle \langle f \mid \alpha_\rho \tilde{f} \rangle. \quad (4.12)$$

Definition 4.2. Let $N \in \mathbb{N}$, $\mathfrak{h} = L^2(M)$ be a Hilbert space, $h = h^*$ a semibounded one-body operator on $\mathfrak{h}$, and $V = V^*$ a pair potential on $\mathfrak{h}$ which is either an infinitesimal perturbation of h or relatively h-bounded with relative bound sufficiently small compared to $1/N$. The **Bogolubov-Hartree-Fock (BHF) energy** $E_{\mathrm{bhf}}(N)$ **(for particle number expectation value $N \in \mathbb{N}$)** is defined by

$$E_{\mathrm{bhf}}(N)$$
$$:= \inf\{\mathrm{Tr}\{\rho\,\mathbb{H}\} \mid \rho \in \mathcal{B}(\mathfrak{F}[\mathfrak{h}]) \text{ quasifree density matrix}, \mathrm{Tr}\{\rho\mathbb{N}\} < \infty\}.$$
$$(4.13)$$

We now assume that $\mathfrak{h} = L^2(\mathbb{R}^2 \times \mathbb{Z}_2)$, that h is a one-body operator which is an infinitesimal perturbation of $-\Delta$ or $\sqrt{-\Delta}$, and that the pair potential $V = V(x, y)$ is given by $\lambda |x - y|^{-1}$. Moreover, we assume ρ to be a quasifree density matrix. In this situation, (3.9) and (4.12) yield

$$\mathrm{Tr}\{\rho\,\mathbb{H}\} = \mathcal{E}_{\mathrm{bhf}}\big(\gamma_\rho^{(1)}, \alpha_\rho\big), \quad (4.14)$$

where the **Bologubov-Hartree-Fock (BHF) energy functional** $\mathcal{E}_{\mathrm{bhf}}$ is defined by

$$\mathcal{E}_{\mathrm{bhf}}(\gamma, \alpha)$$
$$:= \mathrm{Tr}\{h\,\gamma\} + \frac{\lambda}{2} \int \big(\gamma(x,x)\,\gamma(y,y) - |\gamma(x,y)|^2 + |\alpha(x,y)|^2\big) \frac{dx\,dy}{|x - y|},$$
$$(4.15)$$

(with suitably defined Schwartz kernels $\gamma(x,y)$ and $\alpha(x,y)$ and the integration over dx including the summation over spin variables.) Evaluating the infimum in (4.14), we obtain that

$$E_{\mathrm{bhf}}(N) = \inf\big\{\mathcal{E}_{\mathrm{bhf}}(\gamma, \alpha) \mid 0 \le \Gamma[\gamma, \alpha] \le 1, \ \gamma = \gamma^*, \ \alpha = -\alpha^T, \ \mathrm{Tr}\{\gamma\} = N\big\}.$$
$$(4.16)$$

Next we note that if $\Gamma[\gamma, \alpha]$ is a g1-pdm of particle number expectation N, then so is $\Gamma[\gamma, 0]$, because $0 \le \Gamma[\gamma, \alpha] \le 1$ implies $0 \le \gamma \le 1$. Moreover

$$\mathcal{E}_{\mathrm{bhf}}(\gamma, 0) = \mathcal{E}_{\mathrm{hf}}(\gamma). \quad (4.17)$$

BHF = HF Theory in the repulsive case $\lambda > 0$

In this section we study the case of repulsive interaction potentials, $\lambda > 0$. Using (4.15)–(4.17) we obtain

$$\mathcal{E}_{\text{bhf}}(\gamma, \alpha) \geq \mathcal{E}_{\text{bhf}}(\gamma, 0) = \mathcal{E}_{\text{hf}}(\gamma), \tag{4.18}$$

which clearly gives

$$E_{\text{bhf}}(N) = E_{\text{hf}}(N) \tag{4.19}$$

in this case.

BHF Theory and Pairing in the attractive case $\lambda < 0$

In case of attractive interaction potentials, $\lambda < 0$, which are relevant for the study of particles interacting by gravity, it turns out to be important that the fermions under consideration have a spin-$\frac{1}{2}$ degree of freedom. Then the one-particle Hilbert space $\mathfrak{h}$ can be written as a sum $\mathfrak{h} = L^2(\mathbb{R}^3) \oplus L^2(\mathbb{R}^3)$ and the 1-pdm $\gamma = \gamma^*$ and the pairing matrix $\alpha = -\alpha^T$ itself may be viewed as 2×2-matrices, and the g1-pdm $\Gamma = \Gamma^*$ even as a 4×4-matrix, with operator-valued entries each. More specifically, γ, α, and Γ can be written as

$$\gamma =: \begin{pmatrix} \gamma_\uparrow & \gamma_o \\ \gamma_o^* & \gamma_\downarrow \end{pmatrix}, \quad \alpha =: \begin{pmatrix} \alpha_\uparrow & \alpha_o \\ -\alpha_o^T & \alpha_\downarrow \end{pmatrix}, \tag{4.20}$$

and

$$\Gamma[\gamma, \alpha] = \begin{pmatrix} \gamma_\uparrow & \gamma_o & \alpha_\uparrow & \alpha_o \\ \gamma_o^* & \gamma_\downarrow & -\alpha_o^T & \alpha_\downarrow \\ \alpha_\uparrow^* & -\overline{\alpha_o} & 1 - \overline{\gamma_\uparrow} & \overline{\gamma_o} \\ \alpha_o^* & \alpha_\downarrow^* & \gamma_o^T & 1 - \overline{\gamma_\downarrow} \end{pmatrix}. \tag{4.21}$$

Note that the self-adjointness of γ implies that $\gamma_\uparrow = \gamma_\uparrow^*$, $\gamma_\downarrow = \gamma_\downarrow^*$, and the antisymmetry of α, implies that $\alpha_\uparrow = -\alpha_\uparrow^T$, $\alpha_\downarrow = -\alpha_\downarrow^T$, but there is no condition on γ_o nor on α_o. This is the main benefit from the assumption of a spin-$\frac{1}{2}$ degree of freedom.

If $\Gamma = \Gamma[\gamma, \alpha]$ in (4.21) is a g1-pdm of particle number expectation N, then so is

$$\Gamma\left[1 \otimes \hat{\gamma}, \sigma^{(1)} \otimes \sqrt{\hat{\gamma} - \hat{\gamma}^2}\right] = \begin{pmatrix} \hat{\gamma} & 0 & 0 & \sqrt{\hat{\gamma} - \hat{\gamma}^2} \\ 0 & \hat{\gamma} & -\sqrt{\hat{\gamma} - \hat{\gamma}^2} & 0 \\ 0 & -\sqrt{\hat{\gamma} - \hat{\gamma}^2} & 1 - \hat{\gamma} & 0 \\ \sqrt{\hat{\gamma} - \hat{\gamma}^2} & 0 & 0 & 1 - \hat{\gamma} \end{pmatrix},$$

$$\tag{4.22}$$

where $\hat{\gamma} \in \mathcal{B}[L^2(\mathbb{R}^3)]$ is the real, self-adjoint operator

$$\hat{\gamma} := \frac{1}{4}\left(\gamma_\uparrow + \gamma_\downarrow + \gamma_\uparrow^T + \gamma_\downarrow^T\right) \tag{4.23}$$

and $\sigma^{(1)}$ is the first Pauli matrix. In [2] the following estimate was proved, which allows for the elimination of α from the minimization of the BHF energy.

Theorem 4.3.

$$\mathcal{E}_{\text{bhf}}(\gamma, \alpha) \geq \mathcal{E}_{\text{bhf}}\left(\mathbf{1} \otimes \hat{\gamma},\, \sigma^{(1)} \otimes \sqrt{\hat{\gamma} - \hat{\gamma}^2}\gamma\right). \tag{4.24}$$

This estimate proves that the pairing matrix α_{bhf} of a minimizer $\Gamma(\gamma_{\text{bhf}}, \alpha_{\text{bhf}})$ of the BHF energy functional is nonvanishing iff $\gamma_{\text{bhf}} \neq \gamma_{\text{bhf}}^2$ deviates from an orthogonal projection. It can easily be seen that this is not always the case (for instance, if h has purely discrete spectrum and $-\lambda > 0$ is sufficiently small), and to prove that in important physical models, pairing does occur remains a challenge for research.

Acknowledgment

It is a pleasure to thank IMS at NUS, and there especially B.-G. Englert and the IMS staff for the hospitality and the generous support during my stay there. For organizing the inspiring workshop I thank H. Siedentop and the program director, H. Araki.

References

1. V. Bach, Error bound for the Hartree-Fock energy of atoms and molecules, *Commun. Math. Phys.*, 147:527–548, 1992.
2. V. Bach, J. Fröhlich, and L. Jonsson, Bogolubov-Hartree-Fock mean field theory for neutron stars and other systems with attractive interactions, *J. Math. Phys.*, (submitted), 2009.
3. V. Bach, E. H. Lieb, and J. P. Solovej, Generalized Hartree-Fock theory and the Hubbard model, *J. Stat. Phys.*, 76:3–90, 1994.
4. F. Dyson and A. Lenard, Stability of matter I, *J. Math. Phys.*, 8:423–434, 1967.
5. F. Dyson and A. Lenard, Stability of matter II, *J. Math. Phys.*, 9:698–711, 1967.
6. V. Ja. Ivrii and I. M. Sigal, Asymptotics of the ground state energies of large Coulomb systems, *Ann. Math.*, 138(2):243–335, 1993.
7. E. H. Lieb, A lower bound for Coulomb energies, *Phys. Lett.*, 70A:444–446, 1979.

208 V. Bach

8. E. H. Lieb, The number of bound states of one-body Schrödinger operators and the Weyl problem, *Proc. Symp. Pure Math.*, 36:241–252, 1980.
9. E. H. Lieb, Thomas-Fermi and related theories of atoms and molecules, *Rev. Mod. Phys.*, 53:603–641, 1981.
10. E. H. Lieb, Variational principle for many-fermion systems, *Phys. Rev. Lett.*, 46(7):457–459, 1981.
11. E. H. Lieb and S. Oxford, An improved lower bound on the indirect Coulomb energy, *Int. J. Quantum Chem.*, 19:427–439, 1981.
12. Elliott H. Lieb and Walter E. Thirring, Bound for the kinetic energy of fermions which proves the stability of matter, *Phys. Rev. Lett.*, 35(11):687–689, September 1975.
13. M. Reed and B. Simon, *Methods of Modern Mathematical Physics: Analysis of Operators*, volume 4, Academic Press, San Diego, 1978.
14. M. Reed and B. Simon, *Methods of Modern Mathematical Physics: I. Functional Analysis*, volume 1, Academic Press, San Diego, 2nd edition, 1980.
15. M. Reed and B. Simon, *Methods of Modern Mathematical Physics: II. Fourier Analysis and Self-Adjointness*, volume 2, Academic Press, San Diego, 2nd edition, 1980.
16. E. M. Stein and G. Weiss, *Introduction to Fourier Analysis on Euclidean Spaces*, Princeton University Press, Princeton, New Jersey, 2nd edition, 1971.
17. W. Thirring, editor, *The Stability of Matter: From Atoms to Stars – Selecta of Elliott H. Lieb*, Springer, 2nd edition, 1997.